Analysis for Time-to-Event Data under Censoring and Truncation

Analysis for Time-to-Event Data under Censoring and Truncation

Hongsheng Dai, Huan Wang

ELSEVIER

AMSTERDAM • BOSTON • HEIDELBERG • LONDON
NEW YORK • OXFORD • PARIS • SAN DIEGO
SAN FRANCISCO • SINGAPORE • SYDNEY • TOKYO
Academic Press is an imprint of Elsevier

Academic Press is an imprint of Elsevier
32 Jamestown Road, London NW1 7BY, UK
525 B Street, Suite 1800, San Diego, CA 92101-4495, USA
225 Wyman Street, Waltham, MA 02451, USA
The Boulevard, Langford Lane, Kidlington, Oxford OX5 1GB, UK

British Library Cataloguing in Publication Data
A catalogue record for this book is available from the British Library

Library of Congress Cataloging-in-Publication Data
A catalog record for this book is available from the Library of Congress

ISBN: 978-0-12-805480-2

For information on all Academic Press publications visit our website at https://www.elsevier.com/

Publisher: Nikki Levy
Acquisition Editor: Glyn Jones
Editorial Project Manager: Jennifer Pierce
Production Project Manager: Omer Mukthar
Designer: Greg Harris

Typeset by SPi Books and Journals

Contents

Chapter 1
Introduction

1.1 Introduction to the book

1.1.1 Censoring and truncation

In statistics, *survival analysis* is to deal with the analysis for one or more nonnegative event times. Observations of nonnegative event times are usually involved in many different research areas and therefore the topic survival analysis has many different names, for example reliability theory, duration modelling or event history analysis. The data collected in such studies are called survival data, where observations mean the time to the event of interest, for example, in medical research the time from a patient's infection of a particular disease to death or in engineering, the lifetime of a particular item. In summary, time to a particular event is of main interest. In particular, the research interest includes the study of the probability distributions of the event time, how the rate of event occurrence depends on certain risk factors and so on.

In practice, when measuring time to events, we usually have a starting calendar time point and an ending calendar point. The time difference of the two calendar time points is the time to event. The starting calendar time point is usually treated as time 0 for all subjects, although they may actually have different starting calendar time point. In some studies, it has a natural time 0. For example, in drug experiment studies time 0 is just the start of the experiments and usually all subjects start at the same time. In epidemiology studies, time 0 could be the subject infection time of an infectious disease. Due to the long period of survival data collection, biased data or data with missing information are usually observed. For example, event times could be right censored when the study may be stopped before the event actually occurs for a subject. Then we only know that the subject will experience the event after the stopping time (the *censoring* time), i.e. the actual event time value should be greater than the censoring value. Some studies may involve left censoring, for example an measurement of a subject (say the tumor size of a patient) cannot be recorded if it is smaller than a certain threshold. Under left censoring, we only know that the actual observation value should be less than the censoring value. If left censoring and right censoring can occur simultaneously, we will have interval censoring or middle censoring. Various types of censoring have been well studied in the last several decades. Apart from censoring, another important type of missing data in survival analysis is called *truncation*. The main different between truncation and censoring is that truncated data have the selection bias in the sample, but censored data do not have. For censored sample, the observations are randomly selected without any selection bias, but there could be missing information for the observed values. On the other hand, the observations in a truncated data set are selected with certain bias and it is problematic if you use the biased sample to represent the whole population.

The most commonly occurred truncation is left-truncation, which usually involve an entry time point in the study. For example, the time point 0 could be the calendar time that a patient catching a particular disease and research interest is in the time event time T, when the patient is cured. Patients, however, may enter the study (then being recorded) after a while, say time L. This could be because the patient is cured (without any treatment or with some self-treatment), before the date that they planned to go to hospital.

Analysis for Time-to-Event Data under Censoring and Truncation.
http://dx.doi.org/10.1016/B978-0-12-805480-2.50001-0,

In other words, a patient may only go to see a doctor if his/her symptoms last for say L days. Then the patient's information will be recorded if $T \geq L$ (the patient goes to see a doctor). Therefore, the observed event time is randomly larger than the event times in the whole population. The observed data is selected with bias. In summary, left truncation is usually caused by an entry time which makes smaller event times unlikely to be selected.

Right-truncation may occur very often if the study has an end of recruitment time. For example in epidemiology studies, the research interest is the event time T from infection to the development of a particular disease. If there is an end of recruitment calendar time, then a subject will be recorded if it develops the disease by the end of recruitment time. Suppose two subjects infected at the same time, only the one which developed the disease before the end of recruitment time can be recorded. The subject with a longer disease event time cannot be recorded. Therefore the observed data could be biased, i.e. subjects with smaller disease times are more likely to be observed. In fact, such a bias could be very serious in the epidemiology study, since more subjects could be infected towards the end of the study, but among those only the ones with shorter disease time (developing disease before the end of recruitment) are more likely to be recorded; the other ones with longer disease times are less likely to be recruited.

If left-truncation and right-truncation occur simultaneously, we will have interval truncated data. A more complicated scenario is that both censoring and truncation occur. This is because the censoring variable and truncation variables are usually correlated, which makes the analysis difficult. For example the end of recruitment could cause a potential right-truncation and also cause a potential right-censoring, or the censoring time (could be the last follow-up time) may be a certain period of time after the truncation time (the end of recruitment). Another type of biased data is called length-biased data, which actually can be viewed as a special case of truncated data. Some discussions about this are provided in later chapters.

1.1.2 Aim of the book

Research studies in survival analysis under truncation have been very active in the last decade, since in long term survival analysis most data sets are actually collected with selection bias. However, the statistical methodologies for truncated data have not been systematically reviewed. Many existing research works focus on very special cases of truncation or some existing methodologies have limited application due to strong modelling conditions or heavy computational cost. This book aims to provide an overview of recent developments in survival analysis under truncation, especially for bivariate survival analysis.

1.2 Examples

Example 1.1 (Left Truncated and Right Censored Death Times of Psychiatric Patients). A sample of 26 psychiatric patients admitted to the University of Iowa hospital during the years 1935-1948 was reported by Woolson (1981). For each of the 26 patients, the data consist of age at admission, gender, duration of follow-up (in years), and status (death or censoring) at the last follow-up (Table 1.1). To find out if psychiatric patients tend to have shorter survival times, Klein and Moeschberger (2003) compared the mortality of these 26 patients to the standard mortality of residents in Iowa. As shown in Figure 1.1, the survival times in this sample are left truncated since a psychiatric patient must survive long enough to be recruited into the study cohort. An individual who was dead before the admission to hospital cannot be observed. This is an example of univariate survival data with left truncation and right censoring.

Example 1.2 (Left Truncated and Right Censored Death Times of Elderly Residents in a Retirement Community). Hyde (1980) reported data on ages at death of 462 individuals who were in residence of a retirement centre located in California during January 1964 to July 1975. Data consists of gender, age at entry (in months), age of death or leaving the centre (in months), and status (dead or alive). Apart from right censoring, the life lengths in this data are also left truncated because an individual who did not survive

Table 1.1: Survival data for psychiatric inpatients.

Subject	*Gender*	*Status*	*Age at admission*	*Age at exit*	*Length of follow-up*
1	F	1	51	52	1
2	F	1	58	59	1
3	F	1	55	57	2
4	F	1	28	50	22
5	M	0	21	51	30+
6	M	1	19	47	28
7	F	1	25	57	32
8	F	1	48	59	11
9	F	1	47	61	14
10	F	1	25	61	36+
11	F	0	31	62	31+
12	M	0	24	57	33+
13	M	0	25	58	33+
14	F	0	30	67	37+
15	F	0	33	68	35+
16	M	1	36	61	25
17	M	0	30	61	31+
18	M	1	41	63	22
19	F	1	43	69	26
20	F	1	45	69	24
21	F	0	35	65	35+
22	M	0	29	63	34+
23	M	0	35	65	30+
24	M	1	32	67	35
25	F	1	36	76	40
26	M	0	32	71	39+

long enough to enter the centre cannot be observed. In other words, those who died at an early age were excluded from the study. Ignorance of this left truncation can produce biased results.

Example 1.3 (Right Truncated Time to Development of AIDS). Lagakos et al. (1988) reported data on the incubation time for 296 individuals who were infected with AIDS virus and developed AIDS by June 30, 1986. Data consists of infection time (in years) measured from April 1, 1978, and incubation time to development of AIDS, measured from the date of infection. In this data, only individuals who have

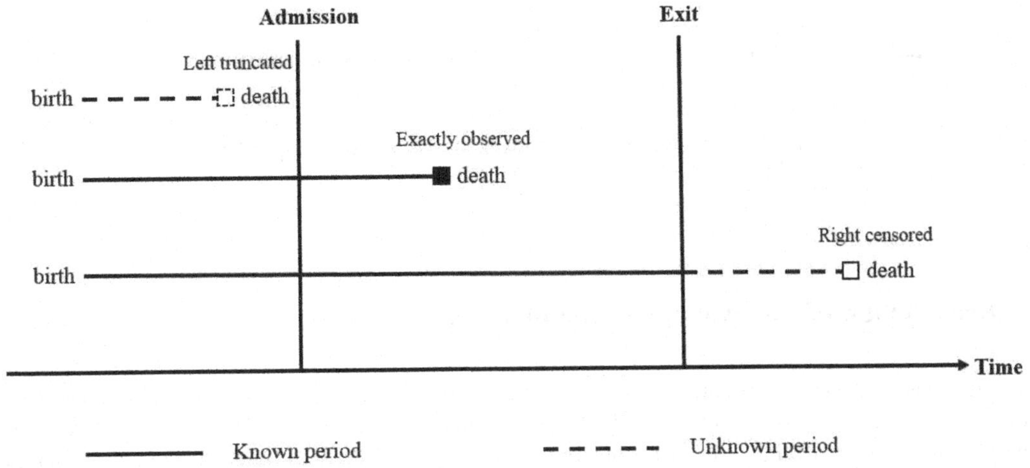

Fig. 1.1: Illustration of left truncated data in Example 1.1.

developed AIDS before June 30, 1986 were recruited. Those with AIDS infection but have yet to develop AIDS before this date were excluded. This type of data is called right truncated survival data.

Example 1.4 (Left Truncated and Right Censored Bivariate Times to Development of AIDS). In a paediatric AIDS cohort study in Shen (2006), the observed pair of event times were the incubation time of mother and the time from birth to development of AIDS for child. A mother-child combination was included only if both were HIV positive and none has developed AIDS. Those with earlier onset of AIDS would then be truncated (not included). After recruitment, the incubation times of mother might be right censored but the time from birth to AIDS development for child could be exactly observed. This is an example where both event times are subject to left truncation and only one of them may be subject to right censoring.

Example 1.5 (Right Truncated and Right Censored Time to Development of Liver Cirrhosis). In a hepatitis C cohort study (Fu et al., 2007; Wang et al., 2013), the event time of interest was the duration from infection with hepatitis C virus to development of liver cirrhosis. The pair of event times observed were the time from infection to recruitment to liver clinic and the time from infection to cirrhosis. Patients with chronic hepatitis C usually have very long incubation period with no or mild symptoms and often seek medical care shortly before development complications, such as cirrhosis. As shown in Figure 1.2, the time to referral is right truncated by the time from infection to the end of recruitment, which means if referral of an individual occurs after the end of 1999, then he/she cannot be included into this cohort, i.e. no information for this individual is available. For those who were recruited, the time from to development of liver cirrhosis may be right censored at the last follow-up time. This is an example of bivariate survival data where one component is right truncated and the other one may be right censored.

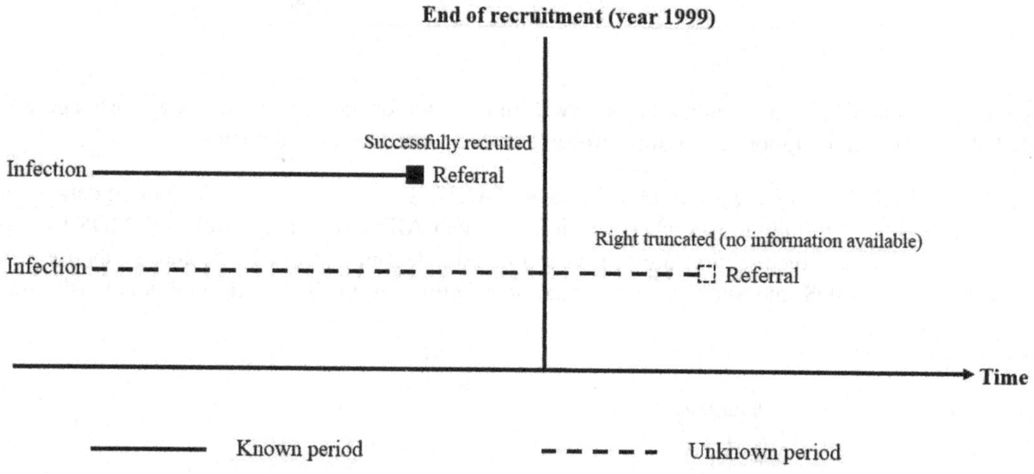

Fig. 1.2: Illustration of right truncated data in Example 1.5.

1.3 Brief review of survival analysis under truncation

There is a vast literature on estimation of distribution or survival function for survival data under truncation. Turnbull (1976) extended the Kaplan-Meier estimator of the survival function in the presence of censoring to incorporate right (or left)-truncation, both fixed and random, but independent of survival. The product limit estimator and its variance for left-truncated data were studied by Woodroofe (1985), Mei-Cheng Wang (1986) and Tsai et al. (1987). Lai and Ying (1991c) proposed a minor modification of the product-limit estimator to estimate a distribution function (either discrete or continuous) in the presence of both left-truncation and right-censoring. They also proved uniform strong consistency and asymptotic normality for their modified estimator. For the left-truncated and right-censored survival data, Wang (1991) derived a new estimator for the distribution of truncation variable and developed the asymptotic properties of the estimator for survival distribution. Their methods, however, were under a more restrictive condition for the truncation distribution. There are some other works related to the univariate survival data under truncation, see for example Keiding and Gill (1990); Gilbels and Wang (1993); He and Yang (2000); Shen (2015); van den Berg and Drepper (2016).

Except for the univariate survival data which has a single event of interest, bivariate survival data which contain pairs of correlated event times are also often observed in medical studies. Estimation of distribution or survival function for bivariate survival data under truncation has received considerable attention. Gürler (1996, 1997) considered the case when only one component of the bivariate data is subject to truncation. Gijbels and Gürler (1998) proposed a useful nonparametric estimator of the bivariate distribution function when one event time is subject to both censoring and truncation and the other one can be fully observed. For bivariate truncated data, van der Laan (1996b) proposed a nonparametric maximum likelihood estimator (NPMLE) of the bivariate survival function, while Huang et al. (2001) proposed a nonparametric estimator of the marginal distributions and applied their estimator to test age-of-onset anticipation in bipolar affective disorder. An inverse-probability-weighted (IPW) approach was proposed by Shen (2006) to estimate the bivariate distribution and survival functions when the paired event times are both left truncated and only of them might be subject to random right censoring. Under certain circumstances, Shen's estimator reduces to the NPMLE proposed by van der Laan (1996b) or the estimator proposed by Gijbels and Gürler (1998). It has been proved that the IPW estimators are consistent under certain conditions and work well for moderate sample sizes. However, the IPW estimators by Shen (2006) require iterative algorithms which may not be practical enough for data with a large sample size. Also, analytic expressions of the asymptotic variances were not presented in his paper. Dai and Fu (2012) considered the bivariate survival data where both event times were subject to random left truncation and random right censoring. A nonparametric estimator for the bivariate survival function was proposed. Their method was based on a polar coordinated transformation which transformed a bivariate survival function to a univariate function. A consistent estimator for the transformed univariate function was proposed, and it could be transformed back to a bivariate form. Their estimator is more computationally efficient than the IPW estimators in Shen (2006), and weakly converges to a Gaussian process with mean zero and an easily estimated covariance function.

Regression models for handling survival data under truncation have also been well studied. P. K. Bhattacharya and Yang (1983) addressed estimation of the parameter in a simple linear regression model, while the univariate event time was assumed to be truncated by a known constant. Lai and Ying (1991c) developed rank regression methods for univariate left-truncated and right-censored data. For accelerated failure time models, Lai and Ying (1994) proposed a general missing information principle and constructed a weighted M-estimator of regression parameter. Gross and Lai (1996) studied AFT models from a view point of curve fitting or approximation theory and proposed another type of M-estimator. He and Yang (2003) considered AFT models for univariate left-truncated data. Using similar idea for the censoring only case in He and Wong (2003), they constructed weighted least squares (WLS) estimator where the weights were random quantities determined by the product limit estimates for the distribution function of truncation variable. Wang et al. (2013) considered AFT models for bivariate survival data under truncation, obtained an unbiased estimate of the regression parameter vector, and developed the asymptotic properties of the proposed estimator.

In the latter chapters, we will introduce some works mentioned above in detail and extend some existing methodologies to more generalised cases of survival analysis under truncation.

1.4 Preliminaries

1.4.1 Lebesgue-Stieltjes Integration

The martingale and counting process theories in this book rely on the Lebesgue-Stieltjes integration, which is briefly introduced in this subsection.

Definition 1.1. A *Borel measure* on $\mathbf{R} = (-\infty, \infty)$ is a nonnegative set function, μ, defined for Borel sets of $\mathbf{R}$ such that

1. $\mu(\phi) = 0$ where ϕ is the empty set;
2. $\mu(I) < \infty$ for each bounded interval I, and
3. $\mu\left(\cup_{i=1}^{\infty} B_i\right) = \sum_{i=1}^{\infty} \mu(B_i)$ for any sequence of disjoint Borel sets $\{B_i, i = 1, 2, \cdots\}$.

The following one-to-one correspondence between a right-continuous increasing function G and a Borel measure μ is the key for the definition of the Lebesgue-Stieltjes integration.

1. $\mu(a,b) = G(b-) - G(a)$;
2. $\mu[a,b] = G(b) - G(a-)$;
3. $\mu[a,b) = G(b-) - G(a-)$;
4. $\mu\{a\} = G(a) - G(a-)$;
5. G is continuous at a if and only if $\mu\{a\} = 0$.

Definition 1.2. Let $f : \mathbf{R} \to \mathbf{R}$ be a Borel measurable function and a right continuous increasing function G and its corresponding Borel measure μ defined above. For any Borel set B, define $\int_B f dG = \int_B f d\mu$.

A simple example for the above definition is when $G : [0, \infty) \to \mathbf{R}$ has derivative g almost everywhere, except on a countable set of discrete jump points, i.e. for points $\{x_1, x_2, \cdots\}$, $\Delta G(x_i) := G(x_i) - G(x_i-) > 0$. Then we can write

$$\int_{[0,t]} f(x) dG(x) = G(0) + \int_0^t f(x) g(x) dx + \sum_i f(x_i) \Delta G(x_i).$$

If we treat ∞ and $-\infty$ as two points and the point sets $\{\infty\}$ and $\{-\infty\}$ as Borel sets, the above function G, Borel measure μ and integration can be defined in the extended real line $\bar{\mathbf{R}} = [-\infty, \infty]$.

Note that throughout this book, $\int_a^b$ means a Riemann integral and $\int_B$ means a Lebesgue-Stieltjes integral.

1.4.2 Probability theory

Throughout this book, we denote Ω as the outcome space of random experiments. An element $\omega \in \Omega$ is a typical outcome and we use $\mathscr{F}$ to denote the σ-algebra (or σ-field) of events and $\mathbf{P}$ to denote a probability measure defined on Ω, where $\mathbf{P}$ satisfies the probability axioms. A probability space is referred to as the triple $(\Omega, \mathscr{F}, \mathbf{P})$.

Definition 1.3. In mathematical analysis and in probability theory, a σ-field $\mathscr{F}$ on Ω is a collection of subsets of Ω, and is closed under countable set operations, i.e. such that

1. $\Omega \in \mathscr{F}$;
2. $A \in \mathscr{F} \Rightarrow A^c \in \mathscr{F}$;
3. $A_n \in \mathscr{F}, n = 1, 2, \cdots, \Rightarrow \cup_{n=1}^{\infty} A_n \in \mathscr{F}$.

Two special σ-fields are $\{\Phi, \Omega\}$, where Φ is the empty set, and $\{A : A \subset X\}$.

Now we can define the random variable and its cumulative distribution function.

Definition 1.4. A random variable X is a function from Ω to the real line $\mathbf{R} = [-\infty, \infty]$. We say that X is *measurable* (relative to $\mathscr{F}$) if $\{X \le x\} := \{\omega : X(\omega) \le x\} \in \mathscr{F}$.

Another example of $\mathscr{F}$ is the σ-field generated by a random variable Y, i.e. $\sigma\{Y\} = \{\{\omega : Y(\omega) \in B\} \subset \Omega, B \in \mathscr{B}(\mathbf{R})\}$ where $\mathscr{B}(\mathbf{R})$ is the Borel σ-field on $\mathbf{R}$.

Definition 1.5. The support of a random variable X (or a vector of random variables $\boldsymbol{X}$) is defined as the set $\{x : \mathbf{P}(X \in (x - r, x + r)) > 0, \forall r > 0\}$ (or $\{\boldsymbol{x} : \mathbf{P}(\boldsymbol{X} \in B(\boldsymbol{x}, r)) > 0, \forall r > 0\}$, where $B(\boldsymbol{x}, r)$ denotes the open ball centered at $\boldsymbol{x}$ with radius r) .

Consider the *cumulative distribution function* of a nonnegative-valued variable X, $F_X(x) = \mathbf{P}(X \le x)$. In practice, we may have $\mathbf{P}(X = \infty) > 0$, i.e. $\mathbf{P}(X < \infty) < 1$. This means the corresponding Borel measure is such that $\mu\{\infty\} > 0$ and $\mu[0, \infty) = \int_{[0,\infty)} dF_X(x) < 1$.

If two random variables X and Y are *independent*, we denote it as $X \perp Y$; if X and Y are independent conditional on Z, we denote $X \perp Y | Z$.

Definition 1.6. Suppose that $\mathscr{G}$ is a sub-σ field of $\mathscr{F}$. The conditional expectation $E(X|\mathscr{G})$ measurable to $\mathscr{G}$ and $E(X|\mathscr{G})$ is such that

$$\int_A E(X|\mathscr{G}) d\mathbf{P} = \int_A X d\mathbf{P}, \quad \forall A \in \mathscr{G}. \tag{1.1}$$

Example 1.6. Suppose $B \in \mathscr{F}$, with $\mathbf{P}(B) > 0$, and $\mathscr{G} = \{\Phi, B, B^c, \Omega\}$. Calculate $\mathbf{P}(A|\mathscr{G})$.

From Definition 1.6, the conditional expectation $\mathbf{P}(A|\mathscr{G}) = E[I_A(\omega)|\mathscr{G}]$ must be measurable to $\mathscr{G}$ and must satisfy

$$\int_C \mathbf{P}(A|\mathscr{G}) d\mathbf{P} = \int_X I_A d\mathbf{P} = \mathbf{P}(A \cap C), \quad \forall C \in \mathscr{G}.$$

Then it is not difficult to guess that $P(A|\mathscr{G}) = aI_B + bI_{B^c}$, for some values of a and b.

Since $C \in \mathscr{G}$, if $C = B$, the above equation tells us

$$\int_C \mathbf{P}(A|\mathscr{G}) d\mathbf{P} = \int_B (aI_B + bI_{B^c}) d\mathbf{P} = a\mathbf{P}(B) = \mathbf{P}(A \cap B)$$

i.e. $a = \mathbf{P}(A|B)$. Similarly, we have $b = \mathbf{P}(A|B^c)$.

Therefore $\mathbf{P}(A|\mathscr{G}) = \mathbf{P}(A|B)I_B + \mathbf{P}(A|B^c)I_{B^c}$.

■

We may also use the following results in this book.

Lemma 1.1. *Suppose X and Y are random variables on $(\Omega, \mathscr{F}, \mathbf{P})$, whose expectations exist. Suppose $\mathscr{G}$ and $\mathscr{H}$ are sub-σ-field of $\mathscr{F}$. Then we have*

1. *If X is measurable with respect to $\mathscr{G}$, then $E(X|\mathscr{G}) = X$;*
2. *If X is independent of $\mathscr{G}$, then $E(X|\mathscr{G}) = E(X)$;*
3. *If $\mathscr{G} \subset \mathscr{H}$, then $E[E(X|\mathscr{G})|\mathscr{H}] = E[X|\mathscr{G}] = E[E(X|\mathscr{H})|\mathscr{G}]$;*
4. *If $X \le Y$, then $E(X|\mathscr{G}) \le E(Y|\mathscr{G})$;*
5. *For constants a, b, $E(aX + bY|\mathscr{G}) = aE(X|\mathscr{G}) + bE(Y|\mathscr{G})$;*
6. *If the expectation of XY exists and X is measurable with respect to $\mathscr{G}$, then $E[XY|\mathscr{G}] = XE[Y|\mathscr{G}]$.*

As in survival analysis events times are usually nonnegative random numbers, the stochastic processes discussed in this book are restricted to processes with nonnegative time indices:

Definition 1.7. A *stochastic process* is a family of random variables $M = \{M(t,\omega), t \in [0,\infty)\}$, defined on $(\Omega, \mathscr{F}, \mathbf{P})$.

Definition 1.8. A family of sub-σ-algebras $\{\mathscr{F}_t : t \geq 0\}$ of $\mathscr{F}$ is called a *filtration* if $s \leq t$ implies $\mathscr{F}_s \subset \mathscr{F}_t$. The right limit of $\mathscr{F}_t$ is defined as $\mathscr{F}_{t+} = \cap_{h>0}\mathscr{F}_{t+h}$; the left limit is defined as $\mathscr{F}_{t-} = \sigma\{\cup_{h>0}\mathscr{F}_{t-h}\}$ or denoted as $\bigvee_{h>0}\mathscr{F}_{t-h}$.

In probability theory, stochastic processes are usually studied on a *stochastic basis*, which is a probability space equipped with a right-continuous filtration, $(\Omega, \mathscr{F}, \mathscr{F}_t, \mathbf{P})$. A typical filtration is $\mathscr{F}_t = \sigma\{M(s) : 0 \leq s \leq t\}$, which represents the information of the process M available up to time t.

Definition 1.9. A stochastic process M is *adapted* to a filtration $\mathscr{F}_t$, if $M(t)$ is $\mathscr{F}_t$-measurable for every t.

The filtration $\mathscr{F}_t = \sigma\{M(s) : 0 \leq s \leq t\}$ is the smallest filtration which the process M is adapted to.

Definition 1.10. The process $M(t)$ is *predictable* if $M(t)$ is $\mathscr{F}_{t-}$ measurable.

The process $M(t)$ is predictable means that $M(t)$ can be determined by its behavior in the past, i.e. $\{M(s), 0 \leq s < t\}$. For a process $M(t)$ the increment over the small interval $[t, t+dt)$ is denoted as $M(dt)$ or $dM(t)$.

Definition 1.11. A *counting process* $N(t)$ is a stochastic process, with $N(0) = 0$ and $N(t) < \infty$, having right-continuous, piecewise constant and only jumps of size 1 and adapted to a filtration $\{\mathscr{F}_t : t \geq 0\}$.

Example 1.7. Suppose X is a nonnegative random variable defined on $(\Omega, \mathscr{F}, \mathbf{P})$. Define the counting process $N(t) = I[X \leq t]$ and $\mathscr{F}_t = \sigma\{N(u), 0 \leq u \leq t\}$. Except the set $\{X \geq t\}$, all other sets in the left limit $\mathscr{F}_{t-}$ are subsets of $\{X < s\}, s \leq t$. Note that $dN(t)$ will be 0 given the set $\{X < t\}$ (or its subset). Therefore we have

$$
\begin{aligned}
E(dN(t)|\mathscr{F}_{t-}) &= I[X \geq t]E[dN(t)|X \geq t] = I[X \geq t]\frac{\mathbf{P}(t \leq X < t+dt)}{\mathbf{P}(X \geq t)} \\
&= I[X \geq t]d\Lambda(t) \qquad (1.2)
\end{aligned}
$$

where $\Lambda(t)$ is called the cumulative hazard function of X. The hazard function $d\Lambda(t)$ is the rate of change of $N(t)$ in the small interval $[t, t+dt)$, conditional on all information available up to time t.

On the other hand, if we equip a more detailed filtration $\mathscr{G}_t$, which is generated from $\mathscr{F}_t$ and $\mathscr{W}_t$ which is based on a covariate process $W(t)$, then

$$
E(dN(t)|\mathscr{G}_{t-}) = I[X \geq t]E[dN(t)|X \geq t, \mathscr{W}_t]. \qquad (1.3)
$$

A more specific model can be assumed for $E[dN(t)|X \geq t, \mathscr{W}_t]$.

Definition 1.12. A *martingale* $M = \{M(t), t \geq 0\}$ is a cadlag (right-continuous with left-hand limit) process, adapted to a filtration $\mathscr{F}_t$, and such that

1. it is integrable, i.e.

$$
\sup_{0 \leq t < \infty} E(|M(t)|) < \infty;
$$

2. it satisfies the martingale property

$$
E(M(t)|\mathscr{F}_s) = M(s), \forall s \leq t.
$$

A martingale $M(t)$ is called a *square integrable* martingale if $\sup_{0\le t<\infty} E(M(t))^2 < \infty$.

Lemma 1.2. *For a right-continuous square integrable martingale $M(t)$, there exists a unique predictable right-continuous increasing process (the predictive variation process) $\langle M,M\rangle$ with $\langle M,M\rangle(0)=0$ and $\langle M,M\rangle(t)<\infty$ for any t, such that $M^2(t)-\langle M,M\rangle(t)$ is a right-continuous martingale.*

The process $\langle M,M\rangle(t)$ is called the predictable variation process, because of $\mathrm{Var}(dM(t)|\mathscr{F}_{t_-}) = d\langle M,M\rangle(t)$. This is from the following facts: $E[dM(t)|\mathscr{F}_{t_-}]=0$,

$$\begin{aligned} d[M^2(t)] &= [M(t_-)+dM(t)]^2-[M(t_-)]^2 \\ &= [2M(t_-)+dM(t)]\cdot dM(t) \\ &= [dM(t)]^2+2M(t_-)dM(t) \end{aligned} \tag{1.4}$$

and

$$\begin{aligned} \mathrm{Var}[dM(t)|\mathscr{F}_{t_-}] &= E[(dM(t))^2|\mathscr{F}_{t_-}] \\ &= E[dM^2(t)|\mathscr{F}_{t_-}]-2E[M(t_-)dM(t)|\mathscr{F}_{t_-}] \\ &= d\langle M,M\rangle(t)-2M(t_-)E[dM(t)|\mathscr{F}_{t_-}] \\ &= d\langle M,M\rangle(t). \end{aligned}$$

If $M(t)$ is a martingale, the process $\int_{[0,t]} H(s)dM(s)$ will also be a martingale, for a bounded and $\mathscr{F}_t$-predictable process $H(t)$.

Theorem 1.1. *If M is martingale defined by $M=N-A$, where N is a counting process and $A(t)=E[N(t)|\mathscr{F}_{t_-}]$, called the compensator, then we have $d\langle M,M\rangle(t)=dA(t)(1-dA(t))$.*

Proof. Integration by parts formula for Lebesgue-Stieltjes integrals (or Equation (1.4)) implies that

$$\begin{aligned} M^2(t) &= \int_{[0,t]} \Delta M(t)dM(t)+2\int_{[0,t]} M(t_-)dM(t) \\ &= \sum_{s\le t}\Delta N(s)[\Delta N(s)-\Delta A(s)]-\int_{[0,t]}\Delta A(s)dM(s)+2\int_{[0,t]}M(t_-)dM(t) \\ &= \sum_{s\le t}\Delta N(s)-\int_{[0,t]}\Delta A(s)dA(s)-2\int_{[0,t]}\Delta A(s)dM(s)+2\int_{[0,t]}M(t_-)dM(t) \\ &= N(t)-\int_{[0,t]}\Delta A(s)dA(s)-2\int_{[0,t]}\Delta A(s)dM(s)+2\int_{[0,t]}M(t_-)dM(t) \\ &= M+\int_{[0,t]}(1-\Delta A(s))dA(s)-2\int_{[0,t]}\Delta A(s)dM(s)+2\int_{[0,t]}M(t_-)dM(t). \end{aligned}$$

The theorem is then proved by noticing that $M-2\int_{[0,t]}\Delta A(s)dM(s)+2\int_{[0,t]}M(t_-)dM(t)$ is a square integrable martingale.

Lemma 1.3. *If M_1 and M_2 are two right-continuous square-integrable martingales with respect to $\mathscr{F}_t$, then there exists a right-continuous predictable process $\langle M_1,M_2\rangle$, called a* predictable covariation pro*cess, with $\langle M_1,M_2\rangle(0)=0$ and $E\langle M_1,M_2\rangle(t)<\infty$, such that $M_1M_2-\langle M_1,M_2\rangle$ is a martingale.*

The process $\langle M_1,M_2\rangle(t)$ is called the predictable covariation process, because of $\mathrm{Cov}(dM_1(t),dM_2(t)|\mathscr{F}_{t_-}) = d\langle M_1,M_2\rangle(t)$. Just like an ordinary covariance,

$$\begin{aligned} \langle aM_1+bM_2,M_3\rangle &= \langle aM_1,M_3\rangle+\langle bM_2,M_3\rangle \\ \langle M_1,M_2\rangle &= \langle M_2,M_1\rangle. \end{aligned}$$

Lemma 1.4. *If M_1 and M_2 are square integrable martingales, H_1 and H_2 are bounded and $\mathscr{F}_t$-predictable processes and $E\left[\int_{[0,t]}H_i(s)dM_i(s)\right]^2<\infty$, then we have*

$$\left\langle \int H_1 dM_1, \int H_2 dM_2 \right\rangle = \int H_1 H_2 d\langle M_1, M_2 \rangle. \tag{1.5}$$

1.4.3 Product integration

In survival analysis, it is often convenient to employ the results for product integrals. The basic concepts and results of product integrals are provided here. We use $\prod$ to denote the product integration sign.

(a)Choose partition $a = t_1 < \cdots < t_k = b$, with $\Delta t = t_j - t_{j-1}$,

$$\prod_{t \in [a,b]} f(t)^{dt} := \lim_{\Delta t \to 0} \prod f(t_i)^{\Delta t} = \exp\left(\int_a^b \log f(t) dt \right)$$

(b)

$$\prod_{t \in [a,b]} [1 + f(t)dt] := \lim_{\Delta t \to 0} \prod [1 + f(t_i)\Delta t] = \exp\left(\int_a^b f(t) dt \right)$$

For functions f and G defined in Definition 1.2 where G is differentiable with a derivative g except at a countable set of points $\{x_1, x_2, \cdots, \}$, we have

$$\prod_{t \in [a,b]} [1 + f(t)dG(t)] := \exp\left(\int_a^b f(t) dG(t) \right) \cdot \prod_{i: x_i \in [a,b]} [1 + f(x_i)\Delta G(x_i)]$$

1.4.4 Right-Censoring

Incomplete observation in the form of right-censoring can rarely be avoided in long-term life history data. The main reason is that the subject is still alive and its lifetime is unknown when the data are collected. Other reasons may be that the subject is lost to follow-up or the subject's death is due to some other reason which is not of interests. For example, if the interest is to study the death from cardiovascular disease, the death event should be censored if one dies from cancer. If we denote T as the event time of interest, our research target is usually to estimate the survival function $S(t) = \mathbf{P}(T > t)$ and the hazard rate function $d\Lambda(t) = \mathbf{P}(t \leq t < t + dt | T \geq t)$. The counting process framework of Example 1.7 can be easily extended to right censoring cases.

Example 1.8. Suppose T and C are independent nonnegative random variables defined on $(\Omega, \mathscr{F}, \mathbf{P})$. Define $X = \min\{T, C\}$ and $\delta = I[T \leq C]$ and the counting process $N(t) = I[X \leq t, \delta = 1]$ and $\mathscr{F}_t = \sigma\{N(u), I[X \leq u, \delta = 0], 0 \leq u \leq t\}$. Except the set $\{X \geq t\}$, all other sets in the left limit $\mathscr{F}_{t-}$ are subsets of $\{X < s\}, s \leq t$. Note that $dN(t)$ will be 0 given the set $\{X < t\}$ (or its subset). Therefore we have

$$\begin{aligned} E(dN(t)|\mathscr{F}_{t-}) &= I[X \geq t] E[dN(t)|X \geq t] = I[X \geq t] \frac{\mathbf{P}(t \leq X < t + dt, \delta = 1)}{\mathbf{P}(X \geq t)} \\ &= I[X \geq t] \frac{\mathbf{P}(t \leq T < t + dt, C \geq T)}{\mathbf{P}(T \geq t, C \geq t)} \\ &= I[X \geq t] \frac{\mathbf{P}(t \leq T < t + dt, C \geq t)}{\mathbf{P}(T \geq t, C \geq t)} \\ &= I[X \geq t] d\Lambda(t). \end{aligned} \tag{1.6}$$

When analysing the censored survival data (X, δ) in Example 1.8, one may assume a particular distribution for T and C, such as exponential distribution with parameter a

$$d\Lambda(t) = adt, \tag{1.7}$$

which implies constant hazard rate overtime.

Other parametric hazard rate function models include Weibull distribution with parameters a and p

$$d\Lambda(t) = a^p p t^{p-1} dt, \tag{1.8}$$

and Gompertz distribution

$$d\Lambda(t) = \exp(\alpha + \beta t)dt. \tag{1.9}$$

These two distribution models incorporate both increasing and decreasing hazard rate functions. Increasing hazard rate functions are appropriate for natural aging or wear, while decreasing hazard rate functions imply that subjects usually have a high failure rate in early stages, such as the failures for certain types of transplants. Sometimes, we may also use bathtub-shaped hazard, for example mortality data may follow this pattern. Higher hazard rate is observed for infants, after which death rate stabilizes, followed by an increasing death rate due to natural aging.

In some research works, people also have observed that the lognormal distribution approximates event times or ages at the onset of certain diseases (Feinleib, 1960; Horner, 1987). Let $\phi(\cdot)$ and $\Phi(\cdot)$ denote the density function and the distribution function of a standard normal distribution. If $\log T$ follows a normal distribution with mean μ and variance σ^2, then the hazard rate function of T is expressed by

$$\lambda(t) = \frac{d\Lambda(t)}{dt} = f(t)/S(t), \tag{1.10}$$

where the density function of T is

$$f(t) = \frac{1}{\sigma t\sqrt{2\pi}} \exp\left[-\frac{1}{2}\left(\frac{\log t - \mu}{\sigma}\right)^2\right] = t^{-1}\phi\left(\frac{\log t - \mu}{\sigma}\right), \tag{1.11}$$

and the survival function of T is

$$S(t) = 1 - \Phi\left(\frac{\log t - \mu}{\sigma}\right). \tag{1.12}$$

The hazard rate of the log normal is hump-shaped, that is, $\lambda(t) = 0$ at $t = 0$, and it increases to a maximum and then decreases to 0 as $t \to \infty$. The problem of this model is that the hazard rate is decreasing for large value of t, which is implausible in many situations.

The gamma distribution has properties similar to the Weibull distributions. Its density function and survival function are

$$f(t) = a^p t^{p-1} \exp(-\lambda t)/\Gamma(p), \quad S(t) = \left[\int_t^{\infty} f(x)d(x)\right] \Big/ \Gamma(p), \tag{1.13}$$

where $a > 0$, $p > 0$, and $\Gamma(\cdot)$ are the gamma function. The hazard rate function, $\lambda(t) = f(t)/S(t)$, is monotone increasing for $p > 1$, with $\lambda(0) = 0$ and $\lambda(t) \to a$ as $t \to \infty$, and monotone decreasing for $p < 1$, with $\lambda(0) = \infty$ and $\lambda(t) \to a$ as $t \to \infty$.

Sometimes we may need to assume a more specific model for $\lambda(t)$.

Example 1.9. (Models for Software Reliability – Example III.1.12 in Andersen et al. (1993))

Let $N(t)$ be the number of software failures in $[0, t]$ and let F be the unknown true number of faults in the program at time 0. When a failure occurs, the corresponding fault is fixed and removed. If we assume that each fault makes the same contribution to the failure rate of the system, the failure rate $\lambda(t)$ is modelled by

$$\lambda(t) = \rho \cdot (F - N(t_-)),$$

where ρ is the true failure rate per fault.

A different model can be assumed as

$$\lambda(t) = \frac{\alpha \cdot (F - N(t_-))}{\beta + t},$$

if we assume that larger faults may tend to produce failures earlier than smaller ones.

However, within survival analysis, data do not simply follow a particular parametric model and it is usually not easy to justify that a parametric distribution fits the data. Therefore nonparametric methods for the distribution estimation of T are preferable in survival analysis. Kaplan and Meier (1958) provided a nonparametric estimator for the survival function of T under right-censoring. Suppose that we have independent observations $X_i, \delta_i, i = 1, \cdots, n$. The Kaplan-Meier estimate for $S(t) = \mathbf{P}(T > t)$ is given by

$$\hat{S}(t) = \prod_{s \leq t} \left\{ 1 - \frac{\Delta N(s)}{Y(s)} \right\}$$

where $\Delta N(s)$ is the number of deaths at time s and $Y(s)$ is the number of subjects which are still at risk at time s.

In many application problems, apart from the censored survival event data, certain important explanatory variables are also known, for example, gender, age, treatment methods and so on. Some explanatory variables may depend on time. For such data, we can use the proportional hazard regression model, where the hazard function is assumed to have the form

$$d\Lambda(t) = d\Lambda_0(t) \exp(\boldsymbol{W}\boldsymbol{\beta})$$

where $\boldsymbol{W}$ is the vector of explanatory variables and $\boldsymbol{\beta}$ is the unknown parameter, which describes the effects of explanatory variables on the hazard rate. For the exponential model in (1.7), a natural extension is to consider

$$d\Lambda(t) = \exp(\boldsymbol{W}\boldsymbol{\beta})dt. \tag{1.14}$$

For the Weibull distribution in (1.8), we may consider the similar model as (1.14) for λ while holding p fixed. Such parametric models assume a particular function form for the baseline hazard rate function $d\Lambda_0(t)$. Under such parametric modelling framework, if we denote f_i as the density function, S_i as the survival function and λ_i and Λ_i as the hazard rate and cumulative hazard rate functions respectively, then one can find the maximum likelihood estimate by maximizing the likelihood function

$$\mathbf{L} \propto \prod_{i \in \mathscr{O}} f(X_i) \cdot \prod_{i \in \mathscr{C}} S(X_i)$$

where $\mathscr{O}$ and $\mathscr{C}$ denote the sets of uncensored and censored observations, respectively, and X_i are the observations. We can further write down the likelihood function as

$$\begin{aligned} \mathbf{L} &\propto \prod_{i=1}^{n} f(X_i)^{\delta_i} S(X_i)^{1-\delta_i} \\ &\propto \prod_{i=1}^{n} \lambda_i(X_i)^{\delta_i} \exp(-\Lambda_i(X_i)). \end{aligned}$$

More generally Λ_0 is usually assumed to have a fully unspecified nonparametric form. Cox (1972) suggested an estimation method for such a proportional hazard model. The idea is that a death event observation, say X_i (with $\delta_i = 1$), will contribute to the likelihood with

$$\frac{\exp(\boldsymbol{W}_i\boldsymbol{\beta})}{\sum_{j\in R(X_i)}\exp(\boldsymbol{W}_j\boldsymbol{\beta})}$$

where $R(X_i) = \{k : X_k \geq X_i\}$. Then the likelihood is defined as the product overall death events, i.e.

$$\prod_i \left[\frac{\exp(\boldsymbol{W}_i\boldsymbol{\beta})}{\sum_{j\in R(X_i)}\exp(\boldsymbol{W}_j\boldsymbol{\beta})}\right]^{I[\delta_i=1]}$$

This partial-likelihood function can also be written in terms of the counting processes $N(t)$, defined in Example 1.8,

$$\prod_{u\in[0,\infty]} \left[\frac{\exp(\boldsymbol{W}_i\boldsymbol{\beta})}{\sum_{j=1}^{n} H_j(u)\exp(\boldsymbol{W}_j\boldsymbol{\beta})}\right]^{dN_i(u)}$$

where $H_j(u) = I[X_j \geq u]$.

As we mentioned before, if death event is due from a disease which is not the interest of the study, such a death event should be treated as censored. However, such censoring variables may be correlated with the main survival event. Therefore independent censoring assumption will not be true. For such cases, it is more reasonable to consider a competing risk model, which can be viewed as a very special Markov transition model (Haller et al., 2013). Considering one transient state 0 (alive) and h absorbing states $k = 1, \cdots, h$. The transition rate, for the ith subject, is modelled by

$$\lambda_{k,i}(t) = \lambda_{k,0}(t)\exp(\boldsymbol{W}_i\boldsymbol{\beta}) \tag{1.15}$$

For such competing risk models, it can be viewed as if there are h event times $T_{i1}, \cdots, T_{ih}$ of interests. We can observe $X_i = \min\{T_{i1}, \cdots, T_{ih}\}$ and the indicator $\delta_i = \{k :$ such that $T_{ik} = X_i\}$.

More general transition models are often used in medical research, for example, the Illness-Death Process model. A patient usually experiences a healthy state 0, a diseased state 1, and a death event 2. A trivariate counting process is usually used, $\boldsymbol{N}(t) = (N_{01}(t), N_{02}(t), N_{12}(t))$, which counts the number of transitions in $[0,t]$. Suppose that the transition rate functions are denoted as $\lambda_{kl,i}$. If λ_{12} only depends on time t, then it is a Markov transition model. One may consider a semi-Markov transition model or even treat the disease event and the death event as a bivariate variable and use nonparametric bivariate analysis to model the transitions. In some long term epidemiology studies, the problem for such Illness-Death Process model is that the data collected could be severely biased due to the biased selection on the illness event time. This book will mainly focus on such issues and use methods based on *truncation* to solve the problem.

Chapter 2
Survival analysis for univariate truncated data

2.1 Introduction

In survival analysis, we usually need to wait very long for occurrence of the events of interest and thus data are often collected over a long time period. For example in hepatitis C cohort study, it may take more than 10 years to observe that a Hepatitis C virus (HCV) infected patient develops cirrhosis (the event of interest). Due to such a long observation period, it is common that incomplete survival data are observed. This is not like other experimental data, which may be collected instantaneously at a certain time point and are observed completely. The most commonly observed incomplete data in survival analysis are censored data. A censored observation means that the observation is only partially known. This is usually because the time to event of interest is very long and at the time that the data are evaluated the event has not happened yet. For example, in a study where the event of interest is development of cirrhosis, if a patient withdrew from the study after 10 years since infection with HCV and has yet developed cirrhosis, we only know that the time to cirrhosis of this individual is longer than 10 years. This is usually referred to right censoring in practice. Other censoring mechanisms include left censoring, interval censoring, middle censoring and so on.

In this chapter, we focus on another type of missing-information scenario, truncation, which is also commonly involved in survival analysis. Truncation occurs when only individuals whose event times are within certain intervals can be observed. Therefore truncation actually introduces sampling bias. Although there is a vast literature for the analysis of truncated data, comparing to censoring, truncation has not been paid enough attention in practice. In most studies, there is often a relatively short window of the duration where the occurrence of the event can be observed. Therefore the collected sample will have potential biases and adjustments to account for the biases are necessary. The following two examples describe the circumstances that truncation occurs in analysis of time to event data.

2.1.1 Basic concepts

Example 2.1. In the data described in Example 1.5, the two event times, time from infection to referral and time from infection to cirrhosis, are correlated because patients with more rapid disease progression are preferentially recruited (Dore et al., 2002). This is because progression of chronic hepatitis C is often asymptomatic after infection. Most infected patients do not seek medical advice until some severe symptoms exhibit. In practice, the closer a patient is to development of cirrhosis, the more likely this individual will go to see a doctor and subsequently be recruited to a liver clinic. In a study based on liver clinics, the HCV patients who have relatively slow disease progression and no severe symptoms by the end of recruitment may not be recruited to liver clinics, i.e. these patients are truncated by the end of recruitment and cannot be observed/included in the study. Therefore, the recruited cohort may have shorter duration from infection to cirrhosis than that of the HCV population (Wang et al., 2013). If so,

Analysis for Time-to-Event Data under Censoring and Truncation.
http://dx.doi.org/10.1016/B978-0-12-805480-2.50002-2,

cohort studies based on liver clinics may overestimate the progression rate among the HCV population. In epidemiology, this is usually called referral bias, which can explain why progression rates to cirrhosis were estimated very differently based on liver clinics assembled by different recruitment methods; see Freeman et al. (2001); Sweeting et al. (2006); Fu et al. (2007) for details.

Example 2.2. In the data described in Example 1.2, truncation occurs when an individual cannot survive long enough to enter the retirement centre. The case that those who died at an early age cannot be observed makes the cohort from the retirement centre be a biased sample when analysing survival experience. Assume that the data collected have the exact life times, say X. However, the sample data are biased because only individuals such that $X > Y$ can be collected, where Y is the time from birth to January 1964, the time point that the program starts. This Y is called the truncation variable and X is left truncated by Y.

Note that truncation is unlike censoring, where an observation may only tell us the event happens within a certain interval. For example, under right censoring, a typical observation for a subject is $(\tilde{X}, \delta)$ with $\tilde{X} = \min\{X, C\}, \delta = I[X \le C]$, where X is the event time and C is the censoring time. Therefore under censoring, there is no sampling bias, only the data are not fully observed.

Definition 2.1. The condition of observing a sample point, for example $X > Y$, is called the *truncation condition.*

Due to the sampling bias under truncation, we should mind our mathematical notations to distinguish the unbiased population and biased samples. We usually use (X,Y) to denote the pair of event time and truncation time. The notation (X,Y) without subscript represent a subject in the whole population. If X and Y are independent, the marginal distributions of X and Y will be of interests. We can denote $F_X(t) = \mathbf{P}(X \le t)$ and $F_Y(t) = \mathbf{P}(Y \le t)$ as the unknown *cumulative distribution functions* for X and Y, respectively. The observed biased sample is denoted as $(X_1,Y_1), (X_2,Y_2), \cdots, (X_n,Y_n)$. Sometimes, we use letters with a $*$, $(X_1^*,Y_1^*), (X_2^*,Y_2^*), \cdots$, to denote the unbiased population. The biased sample $(X_1,Y_1), (X_2,Y_2), \cdots, (X_n,Y_n)$ is a subsequence of $(X_1^*,Y_1^*), (X_2^*,Y_2^*), \cdots$. Therefore we have, for any $i = 1, \cdots, n$,

$$\mathbf{P}(X_i \le t) = \mathbf{P}(X \le t | X > Y) = \mathbf{P}(X_i^* \le t | X_i^* > Y_i^*). \tag{2.1}$$

Note that this is unlike the case of unbiased samples, where the distribution for a random observation is the same as the population distribution.

We assume that X is independent of Y. However, the observed pair X_i and Y_i are not independent, since the observed data are biased and the truncation condition $X_i > Y_i$ for observing the pair (X_i, Y_i) forces a correlation between them. The probability $\mathbf{P}(X > Y)$ is called *truncation probability* and we assume throughout this book: $\mathbf{P}(X > Y) > 0$.

Note that some existing research works, such as He and Yang (1998) and He and Yang (2000), use the truncation condition $X \ge Y$. The methodologies introduced in this chapter are also applicable under the truncation condition $X \ge Y$, if the following Condition 2.1.1 holds.

Condition 2.1.1 $\mathbf{P}(X = Y) = 0$.

If the following Condition 2.1.2 holds, the methodologies introduced in this chapter are applicable only under the truncation condition $X > Y$.

Condition 2.1.2 $\mathbf{P}(X = Y) > 0$.

Condition 2.1.2 means that the distributions of X and Y are discrete and they assign positive probabilities for some common discrete points. For such scenario, if we use the *revised data* method introduced in He and Yang (1998), the martingale methods in this chapter could possibly be generalised to cases under the truncation condition $X \ge Y$ and Condition 2.1.2. For simplicity, this methodology is not covered here and we use the truncation condition $X > Y$ throughout the book.

Other types of truncation

We can define right truncation and interval truncation similarly as left truncation. In fact, for the data in Examples 1.2 and 2.2 if the program ends in July 1975, there could be a right truncation involved either. Therefore this example could also be interpreted as interval truncation.

Truncation can be viewed as a selection bias. Another type of selection bias, named length-bias, is often used in biological studies. For example, when an observer moves along a selected path collecting samples of animals, insects or plants, individuals that are closer to the observer's path have higher probability of being detected. In summary, the distribution of observations is length-biased if the observations are selected with probability proportional to their lengths.

Formally, a distribution F^* is called a length-biased distribution, corresponding to a distribution F (both distributions are defined on $\mathbf{R}^+$), if

$$F^*(x) = c^{-1} \int_0^\infty t dF(t), \tag{2.2}$$

where $c = \int_0^\infty t dF(t)$. We may also denote the relation in (2.2) as

$$dF^*(x) = c^{-1} x dF(x), \tag{2.3}$$

which means that the biased distribution is proportional to the unbiased distribution with a factor x. This is actually a special case of the truncation discussed in this book, since for example under left (independent) truncation (only observing X with $X > Y$) we have

$$\mathbf{P}(X_i \in dx) = \frac{\mathbf{P}(X \in dx)\mathbf{P}(x > Y)}{\mathbf{P}(X > Y)}, \tag{2.4}$$

which implies that the biased distribution is proportional to the unbiased distribution with a factor $\mathbf{P}(x > Y)/\mathbf{P}(X > Y)$. Although there are many different approaches to deal with length biased sampling problems, truncation models are appropriate for many scenarios.

2.1.2 Survival functions and product limit integration

In this subsection, we consider the homogeneous case, where no risk factors are involved and all subjects in the whole population are identically and independently distributed (i.i.d.). We focus on the estimation of $F_X(t)$, which is the cumulative distribution function of X.

In practice, it is often convenient to discuss the distribution of the event time X via its hazard rate function. Suppose that X is a continuous random variable with density function $f_X(t) = dF_X(t)/dt$. Then its *hazard rate function* is defined as

$$\begin{aligned}
\lambda_X(t) &= \lim_{\Delta t \to 0} \frac{\mathbf{P}(t \le X < t + \Delta t | X \ge t)}{\Delta t} \\
&= -\left[\frac{d}{dt} S_X(t)\right] / S_X(t) \\
&= \frac{f_X(t)}{S_X(t)},
\end{aligned} \tag{2.5}$$

where $S_X(t) = 1 - F_X(t)$ is called the *survival function*. Estimation of the distribution function F_X can be based on the estimation of S_X or the *cumulative hazard function* $\Lambda_X(t) = \int_0^t \lambda_X(t)dt$, since Λ_X and S_X are uniquely determined by each other through the relationship $S_X(t) = \exp(-\Lambda_X(t))$. As the random variable X may not be continuous, it is more appropriate to use the following general notation,

$$\begin{aligned}\Lambda_X(t) &= \int_{[0,t]} \mathbf{P}(t \le X < t+dt | X \ge t) \\ &= \int_{[0,t]} \mathbf{P}(t \le X < t+dt)/S_X(t-) \\ &= \int_{[0,t]} \frac{dF_X(t)}{S_X(t-)} \end{aligned} \tag{2.6}$$

where the integration above is the Lebesgue-Stieltjes integration. Further, we have $d\Lambda_X(t) = \frac{dF_X(t)}{S_X(t-)}$. If F_X has the derivative f_X except in a countable set, say B, of nondifferentiable points, we can write $S_X(t) = \exp(-\int_0^t f_X(t)/S_X(t)dt) \cdot \prod_{t \in B}(1 - \Delta F_x(t)/S_X(t-))$.

Consider a random variable X with $a_{F_X} := \inf\{t : \mathbf{P}(X \le t) > 0\}$ and $b_{F_X} := \sup\{t : \mathbf{P}(X \le t) < 1\}$, where $b_{F_X} \in (0, \infty]$. If the survival function S_X is continuous at b_{F_X}, then $\Lambda_X(b_{F_X}) = \infty$ is well defined in (2.6). Note that when $b_{F_X} = \infty$, S_X being continuous at $b_{F_X} = \infty$ means that $S_X(b_{F_X}) = 0$, i.e. $\mathbf{P}(X = \infty) = 0$.

If the survival function S_X (or the distribution function F_X) has a jump at b_{F_X} i.e. $S_X(b_{F_X}) = 0$ but $S_X(b_{F_X}-) > 0$, then $\Delta\Lambda_X(b_{F_X}) = \mathbf{P}(X = b_{F_X} | X \ge b_{F_X}) = 1$, regardless of how big the jump $\Delta F_X(b_{F_X})$ is. This also implies that $\Lambda_X(b_{F_X}) < \infty$ since $S_X(b_{F_X}-) > 0$ and $\Lambda_X(b_{F_X}-) < \infty$. The survival function S_X has a jump at b_{F_X} is quite common in practice, especially when $b_{F_X} = \infty$. This means that $\mathbf{P}(X = \infty) > 0$, which indicates that there is a positive probability that the event will never happen (for example if someone is immune to an infectious disease, the infection will never happen). For such problems, we usually model the survival function $S_X(t)$ (or hazard function $\Lambda_X(t)$) for $t \in [0, b_{F_X})$ and the probability $\mathbf{P}(X = \infty)$ separately, via the so-called *cure models*. Note that in practice, we usually can only consider a time interval $[a_{F_X}, \tau]$, $\tau \le b_{F_X}$, since the practical experiment may be followed only up to time τ (or the maximum censoring time). The cure model will be more appropriate, since F_X cannot be identified in $(\tau, b_{F_X}]$ but the probability $S_X(\tau)$ can be modelled via cure models. We will discuss the cure models at the end of this chapter, but before that, we will introduce the following conditions and will refer to these conditions wherever it is used.

Condition 2.1.3 *1. The random variable X is such that its survival function S_X is continuous at τ, the maximum time point under consideration;*
2. The survival function of random variable X is continuous at its left ending point, a_{F_X}.

Note that the item 2 in the above condition is for the continuity of S_X at the left ending point, we may need it in some occasions because of the left truncation problem considered in the book.

Equation (2.6) motivates the product-limit estimator for the survival function, which is shown in the following example.

Example 2.3. Consider an i.i.d. sample $(\tilde{X}_1, Y_1, \delta_1), (\tilde{X}_2, Y_2, \delta_2), \cdots, (\tilde{X}_n, Y_n, \delta_n)$ generated on a probability space $(\Omega, \mathscr{F}, \mathbf{P})$, under right censoring and left truncation, where Y_i is the left truncation variable and $\tilde{X}_i = \min\{X_i, C_i\}$ is the observed event time and $\delta_i = I[X_i \le C_i]$ is the censoring indicator. Note that for such left truncated and right censored data, there are two different truncation conditions.

Condition 2.1.4 *Data can only be observed if $X > Y$ and $\mathbf{P}(C > Y) = 1$ (Shen, 2006).*

Condition 2.1.5 *Data can only be observed if $\tilde{X} = \min\{X, C\} > Y$.*

Condition 2.1.4 may be too strong to be true in practice as it requires $C > Y$. For example, Y may be the enrolment time of a subject, X can be the time to a disease of interest and the censoring time C may be the time of death (due to other unrelated reasons) of the subject. It is likely that $C < Y$ and such subjects will not be observed. Therefore, the observed Y is actually subject to truncation by $Y < \min\{X, C\}$. Therefore Condition 2.1.5 is more reasonable in practice. This book mainly consider the more general Condition 2.1.5. It will be clearly stated if any method needs the stronger Condition 2.1.4.

If we assume that (Y, C) is independent of X (but Y and C can be correlated), then under Condition 2.1.4,

$$
\begin{aligned}
& \frac{\mathbf{P}(t \leq \tilde{X}_i < t+dt, \delta_i = 1)}{\mathbf{P}(\tilde{X}_i \geq t > Y_i)} \\
= & \frac{\mathbf{P}(t \leq \tilde{X} < t+dt, \delta = 1 | X > Y)}{\mathbf{P}(\tilde{X} \geq t > Y | X > Y)} \\
= & \frac{\mathbf{P}(t \leq X < t+dt, C \geq t | X > Y)}{\mathbf{P}(X \geq t, C \geq t, t > Y | X > Y)} \\
= & \frac{\mathbf{P}(t \leq X < t+dt, C \geq t, t > Y)/\mathbf{P}(X > Y)}{\mathbf{P}(X \geq t, C \geq t, t > Y)/\mathbf{P}(X > Y)} \\
= & \frac{\mathbf{P}(t \leq X < t+dt)}{\mathbf{P}(X \geq t)} = d\Lambda_X(t). \qquad (2.7)
\end{aligned}
$$

Note that equation (2.7) is also true under Condition 2.1.5 (to prove it, simply replace the condition $X > Y$ by $\tilde{X} > Y$).

Therefore, a straightforward estimator for $d\Lambda_X(t)$ is given by

$$
d\hat{\Lambda}_X(t) = \frac{n^{-1}\sum_{i=1}^n I[t \leq \tilde{X}_i < t+dt, \delta_i = 1]}{n^{-1}\sum_{i=1}^n I[\tilde{X}_i \geq t > Y_i]}, \qquad (2.8)
$$

where the numerator represents the number of subjects who failed at time t and the denominator represents the number of subjects at risk. Then by noticing that $S(t)$ can be written as a product integration

$$
S_X(t) = \prod_{s \in [0,t]} [1 - d\Lambda_X(s)], \qquad (2.9)
$$

we have that the product-limit estimator (Kaplan-Meier estimator if no truncation is involved) is given by

$$
\hat{S}_X(t) = \prod_{s \in (0,t]} \left[1 - d\hat{\Lambda}_X(s)\right]. \qquad (2.10)
$$

■

The statistical properties of the above estimator can be derived via martingale theories.

2.1.3 *Preliminaries and the martingale theories*

Although details of the application martingale theories on survival analysis were provided in Fleming and Harrington (1991) and Andersen et al. (1993), the methodologies in their book are mainly based on censored data. We provide a brief introduction about the martingale theories for truncated and censored data and emphasize the difference of the methodologies between truncation and censoring.

In this section we demonstrate the theory based on the context of Example 2.3. For a cumulative distribution function $F(x)$, we denote $a_F := \inf\{x : F(x) > 0\}$ and $b_F := \sup\{x : F(x) < 1\}$. For simplicity, we always impose the following assumption on a random variable X.

Condition 2.1.6 *Continuous support region: the support of* X *is* $[a_{F_X}, b_{F_X}]$.

Note that, although Condition 2.1.6 rules out the cases with $\mathbf{P}(X \in A) = 0$ for some interval A (with positive Lebesgue measure) such that $A \subset (a_{F_X}, b_{F_X})$, the methodologies developed in this book are still valid even if the support is not continuous.

For the function $\mathbf{P}(C \geq t > Y)$, we denote $a := \inf\{t : \mathbf{P}(C \geq t > Y) > 0\}$ and $b := \sup\{t : \mathbf{P}(C \geq t > Y) > 0\}$. Denote $\tau := b_{F_X} \wedge b$, where $\tau \in (0, \infty]$ is the terminal time (the end of the time period under consideration).

Some existing research works assumes

Condition 2.1.7 $a_{F_Y} \leq a_{F_X} < b_{F_C}$.

The condition $a_{F_Y} \leq a_{F_X}$ implies that there is a chance to observe all values of X. If $a_{F_Y} > a_{F_X}$ then X cannot be observed when $X \in [a_{F_X}, a_{F_Y}]$. The condition $a_{F_X} < b_{F_C}$ implies not all values of X are censored.

This condition however is not strong enough to guarantee the identifiability of F_X, when Y and C are correlated. For example, if $Y = C$, even if Condition 2.1.7 is true we still cannot observe any uncensored X. Thus, in this book we consider the following requirement.

Condition 2.1.8 *$a \leq a_{F_X}$ and the function $\mathbf{P}(Y < t \leq C) > 0$ (as a function of t), almost surely with respect to dF_X, in the interval $[a_{F_X}, \tau]$.*

Condition 2.1.8 implies $\mathbf{P}(X > Y) > 0$.

As we focus on event times with nonnegative values, for simplicity, we assume that

Assumption 2.1.1 $a = 0$,

and therefore, we consider a fixed time interval $[0, \tau]$. Note that if $a_{F_X} = 0$, we will need item 2 in Condition 2.1.3, i.e. F_X is continuous at $a_{F_X} = 0$; otherwise if F_X assigns an atom mass at 0 then this atom mass cannot be identified due to the truncation condition $X > Y$.

Also note that Condition 2.1.8 specifies a requirement for the function $\mathbf{P}(Y < t \leq C)$. It can be simplified under Condition 2.1.4 ($\mathbf{P}(Y < C) = 1$),

$$\mathbf{P}(Y < t \leq C) = \mathbf{P}(t \leq C) - \mathbf{P}(t \leq C, t \leq Y) = \mathbf{P}(t \leq C) - \mathbf{P}(t \leq Y). \tag{2.11}$$

Therefore, under Condition 2.1.4, we may impose a condition for $\mathbf{P}(t \leq C)$ and $\mathbf{P}(t \leq Y)$. We here impose the above condition on $\mathbf{P}(Y < t \leq C)$ since we are considering the more general Condition 2.1.5, under which (2.11) is not satisfied.

We denote the counting processes

$$N_i(t) = I[\tilde{X}_i \leq t, \delta_i = 1], \;\; N_i^C(t) = I[\tilde{X}_i \leq t, \delta_i = 0]. \tag{2.12}$$

Define the filtration

$$\mathscr{F}_{i,t} = \sigma\{N_i(u), N_i^C(u), I[Y_i \leq u] : 0 \leq u \leq t]\}, \;\; t \in [0, \tau] \tag{2.13}$$

and its left limit is $\mathscr{F}_{i,t-} = \sigma\{N_i(u), N_i^C(u), I[Y_i \leq u] : 0 \leq u < t]\}$.

Lemma 2.1. *With the above notation, we have*

$$E\{dN_i(t)|\mathscr{F}_{i,t-}\} = I[\tilde{X}_i \geq t > Y_i]\Lambda_X(dt) := A_i(dt). \tag{2.14}$$

Proof. We provide the proof under Condition 2.1.4. The proof under Condition 2.1.5 follows easily.

From (1.1), we need to show that for any set $A \in \mathscr{F}_{i,t-}$,

$$\int_A I[\tilde{X}_i \geq t > Y_i]\Lambda_X(dt)d\mathbf{P} = \int_A dN_i(t)d\mathbf{P}. \tag{2.15}$$

or

$$\mathbf{P}(\{\tilde{X}_i \geq t > Y_i\} \cap A) \cdot \Lambda_X(dt) = \mathbf{P}(\{dN_i(t) = 1\} \cap A). \tag{2.16}$$

(a) A set $A \in \mathscr{F}_{i,t-}$, in the form of $\{\tilde{X}_i \geq s_3, Y_i \in [s_1, s_2)\}$ for any $0 \leq s_1 < s_2 < t, s_3 \leq t$, guarantees (2.16) since it becomes

$$\mathbf{P}(\tilde{X}_i \geq t, Y_i \in [s_1, s_2)) \cdot \Lambda_X(dt) = \mathbf{P}(\tilde{X}_i \in [t, t+dt), \delta_i = 1, Y_i \in [s_1, s_2))$$

or equivalently

$$\mathbf{P}(\tilde{X} \geq t, Y \in [s_1, s_2) | X > Y) \cdot \Lambda_X(dt) = \mathbf{P}(\tilde{X} \in [t, t+dt), \delta = 1, Y \in [s_1, s_2) | X > Y), \tag{2.17}$$

which is obviously true since both sides are

$$\frac{\mathbf{P}(X \geq t)\mathbf{P}(C \geq t, Y \in [s_1, s_2))}{\mathbf{P}(X > Y)} \cdot \Lambda_X(dt) = \frac{\mathbf{P}(t \leq X < t+dt)\mathbf{P}(C \geq t, Y \in [s_1, s_2))}{\mathbf{P}(X > Y)}.$$

(b) A set $A \in \mathscr{F}_{i,t-}$, in the form of $\{\tilde{X}_i \in [s_3, s_4), \delta_i = 1, Y_i \in [s_1, s_2)\}$ (or $\{\tilde{X}_i \in [s_3, s_4), \delta_i = 0, Y_i \in [s_1, s_2)\}$) for any $0 \leq s_3 < s_4 \leq t$ and any s_1, s_2 such that $A \in \mathscr{F}_{i,t-}$, also guarantees (2.16) since both sides of the equation become 0.

(c) A set $A \in \mathscr{F}_{i,t-}$, in the form of $\{\tilde{X}_i \geq s_3, Y_i \geq s_1\}$ for any $0 \leq s_1 \leq t, 0 \leq s_3 \leq t$, guarantees (2.16) since it becomes

$$\mathbf{P}(\tilde{X}_i \geq t > Y_i \geq s_1) \cdot \Lambda_X(dt) = \mathbf{P}(\tilde{X}_i \in [t, t+dt), \delta_i = 1, Y_i \geq s_1) \tag{2.18}$$

and further

$$\frac{\mathbf{P}(\tilde{X} \geq t, t > Y \geq s_1)}{\mathbf{P}(X > Y)} \cdot \Lambda_X(dt) = \frac{\mathbf{P}(\tilde{X} \in [t, t+dt), \delta = 1, t > Y \geq s_1)}{\mathbf{P}(X > Y)} \tag{2.19}$$

which is also true.

For any other sets, which are union, intersection, complementary of the sets of form in the above items **(a)**, **(b)** and **(c)**, equation (2.16) also holds. ■

Note that the above proof uses the condition that X and Y are independent. If they are not independent, more complicated models should be introduced in order to derive a similar result as Lemma 2.1. This is beyond the aim of this book and we leave it to future research work. Also note that in the above proof, equation (2.15) will not be true if the truncation condition is $X \geq Y$ and $\mathbf{P}(X = Y) > 0$ (under Condition 2.1.2). This is easily seen since for a set $A = \{\tilde{X}_i \geq t, Y_i \geq t\} \in \mathscr{F}_{i,t-}$, the left-hand side of equation (2.15) is 0, the right-hand side, however, is not 0 under Condition 2.1.2.

If we denote

$$M_i(t) = N_i(t) - A_i(t) \tag{2.20}$$

we have $E\{dM_i(t)|\mathscr{F}_{i,t-}\} = 0$ since $A_i(t)$ is adaptive to $\mathscr{F}_{i,t}$, i.e. $M_i(t)$ is a martingale. The predictive quadratic variation of $M_i(t)$ is denoted as $\langle M_i, M_i \rangle(t)$ and we have $\langle M_i, M_i \rangle(dt) = A_i(dt)(1 - A_i(dt))$.

If we consider n i.i.d. observations in Example 2.3 and define

$$M(t) = \sum_{i=1}^{n} M_i(t) = N(t) - A(t), \tag{2.21}$$

with

$$N(t) = \sum_{i=1}^{n} N_i(t),$$
$$A(t) = \sum_{i=1}^{n} A_i(t), \qquad (2.22)$$

it can be shown that $M(t)$ is a $\mathscr{F}_t := \sigma\{\mathscr{F}_{i,t}, i = 1, \cdots, n\}$-martingale with predictive quadratic variation $\langle M, M\rangle(t)$,

$$\langle M, M\rangle(dt) = H(t) \cdot \Lambda_X(dt)(1 - \Lambda_X(dt)) \qquad (2.23)$$
$$H(t) = \sum_{i=1}^{n} I[\tilde{X}_i \geq t > Y_i]$$

Suppose that $\mathbf{H}(s)$ is a bounded predictable $\mathscr{F}_t$-process, depending on the sample size n. Define $\mathscr{L} = \{t : \mathbf{P}(C \geq t > Y)S_X(t-) > 0\}$, $\bar{t} = \sup \mathscr{L}$ and $\underline{t} = \inf \mathscr{L}$. Note that we assume that Assumption 2.1.1 is true in the remaining parts of this section, under which we have $\underline{t} = 0$. Clearly $\underline{t} = 0 \notin \mathscr{L}$ since Y is nonnegative.

Condition 2.1.9 *(a)For any $t \in \mathscr{L}$, $\Lambda_X(t) < \infty$;*
(b)There exists a non-negative and left-continuous function with right-hand limits, $h(s)$, such that

$$\sup_{\varepsilon \leq s \leq t} \left| [\mathbf{H}(s)]^2 H(s) - h(s) \right| \xrightarrow{P} 0, \quad n \to \infty,$$

and $h(s)$ is bounded in $[\varepsilon, t]$, whenever $\varepsilon, t \in \mathscr{L}$ and $h(s) = 0, \forall s \notin \mathscr{L}$;
(c)$\int_{\mathscr{L}} h(s)(1 - \Delta\Lambda(s))d\Lambda_X(s) < \infty$, when $\bar{t} \notin \mathscr{L}$;
(d)$\lim_{\varepsilon\downarrow 0} \limsup_{n\to\infty} P\left\{\int_{[0,\varepsilon]} [\mathbf{H}(s)]^2 H(s) d\Lambda_X(s) > \xi\right\} = 0, \forall \xi > 0$;
(e)$\lim_{t\uparrow\tau} \limsup_{n\to\infty} P\left\{\int_{[t,\tau]} [\mathbf{H}(s)]^2 H(s) d\Lambda_X(s) > \xi\right\} = 0, \forall \xi > 0$, when $\bar{t} \notin \mathscr{L}$.

The next theorem, following from Theorem 6.2.1, Fleming and Harrington (1991), provides the weak convergence results of stochastic integrals with respect to the martingale M.

Theorem 2.1. *Let $Z(t)$ be a zero-mean Gaussian process with continuous sample paths, independent increments and variance function $\int_{[0,t]} h(s)(1 - \Delta\Lambda_X(s))d\Lambda_X(s)$ and denote*

$$\mathcal{M}(t) = \int_{[0,t]} \mathbf{H}(s) dM(s).$$

Under parts (a) and (b) of Condition 2.1.9 we have that

$$\mathcal{M} \Rightarrow Z,$$

on $D[\varepsilon, t]$, for any $0 < \varepsilon < t < \bar{t}$, as $n \to \infty$, where D and $\Rightarrow$ are defined in Definition 2.2 and Definition 2.3, respectively.

Under parts (a), (b), (c) and (d) of Condition 2.1.9 we have that

$$\mathcal{M} \Rightarrow Z,$$

on $D[0, t]$, for any $0 < t < \bar{t}$, as $n \to \infty$.

Under parts (a), (b), (c), (d) and (e) of Condition 2.1.9 we have that

$$\mathcal{M} \Rightarrow Z,$$

on $D[0, \bar{t}]$, as $n \to \infty$.

Note that in the above theorem, $\bar{t}$ can be ∞.

Definition 2.2. We denote $D[0,b]$ as the set of all functions from $[0,b]$ to the real axis $\mathbf{R}$ that are right continuous with left-limit. The set $D[a,b]$ becomes a metric space if it is equipped with the metric $||\cdot||$, for $m_1, m_2 \in D[0,b]$,

$$||m_1 - m_2|| = \sup_{0\le t\le b} |m_1(t) - m_2(t)|. \tag{2.24}$$

We denote the Borel σ-field of subsets of $D[0,b]$ as $\mathscr{D}$, which is generated by the open balls.

The weak convergence $\Rightarrow$ in Theorem 2.1 is defined as follows.

Definition 2.3. Consider stochastic processes M_n, $n \ge 0$ defined on the common probability space $(\Omega, \mathscr{F}, \mathbf{P})$, with $M_n(\omega)$ as a left-limit and right-continuous function on $[0,b]$, i.e. $M_n(\omega) \in D[0,b]$. Denote $\mathbf{Q}_n$ as the probability measure induced on $(D[0,b], \mathscr{D})$ by M_n. We say that M_n weakly converges to M_0 (or $\mathbf{Q}_n$ weakly converges to $\mathbf{Q}_0$), if $\int g d\mathbf{Q}_n \to \int g d\mathbf{Q}$ for all real value function g on $D[0,b]$ that are bounded, uniformly continuous and $\mathscr{D}$-measurable.

The weak convergence is denoted by $M_n \Rightarrow M_0$ on $D[0,b]$.

2.2 Nonparametric estimation

2.2.1 Estimation for the cumulative hazard

In this section, we will study how to use the martingale techniques to develop the large sample properties for $\hat{\Lambda}_X(t)$ given in (2.8). From the following result,

$$\begin{aligned}\hat{\Lambda}_X(t) - \Lambda_X(t) &= \int_{[0,t]} \left\{ \frac{N(ds)}{H(s)} - \Lambda_X(ds) \right\} \\ &= \int_{[0,t]} \frac{I[H(s)>0]}{H(s)} M(ds) - \int_{[0,t]} I[H(s)=0] \Lambda_X(ds)\end{aligned} \tag{2.25}$$

we know that $\hat{\Lambda}_X(t)$ is a biased estimate to $\Lambda_X(t)$, but the bias $\int_{[0,t]} I[H(s)=0]\Lambda_X(ds)$ (even with a multiple factor n) is negligible (Fleming and Harrington, 1991). The variance of $\hat{\Lambda}_X(t)$ is

$$\begin{aligned}Var(\hat{\Lambda}_X(t)) &= E\left\{ \hat{\Lambda}_X(t) - \Lambda_X(t) + \int_0^t I[H(s)=0]\Lambda_X(ds) \right\}^2 \\ &\approx E\left\{ \int_0^t \frac{I[H(s)>0]}{H(s)} M(ds) \right\}^2 \\ &= \int_{[0,t]} E\left\{ \frac{I[H(s)>0]}{H(s)} \right\} (1-\Lambda_X(ds))\Lambda_X(ds)\end{aligned} \tag{2.26}$$

Under the following assumption,

Assumption 2.2.1 *[1.] $\Lambda_X(t) < \infty$ for $t \in [0,\tau)$, and $\int_{(0,\tau)} [\mathbf{P}(C \ge s > Y) \cdot S_X(s-)]^{-1} d\Lambda_X(s) < \infty$ (note that this assumption implies Condition 2.1.8 is true);*
[2.] Item 2 of Condition 2.1.3 is satisfied, which implies that $\Lambda_X(t)$ is continuous at $t=0$;
[3.] Item 1 of Condition 2.1.3 is satisfied, which implies that $\Lambda_X(t)$ is continuous at $t=\tau$;

we have the following lemma and theorem.

Lemma 2.2. *Under Condition 2.1.6 and Condition 2.1.8, we have the following statements: for any $\kappa \in (0,1)$ we have that*

I. with item [1.] of Assumption 2.2.1,

$$n^{1-\kappa}\int_{(0,\tau)} I[H(s)=0]\Lambda_X(ds) \xrightarrow{P} 0; \tag{2.27}$$

II. with items [1.] and [2.] of Assumption 2.2.1,

$$n^{1-\kappa}\int_{[0,\tau)} I[H(s)=0]\Lambda_X(ds) \xrightarrow{P} 0; \tag{2.28}$$

III.with items [1.], [2.] and [3.] of Assumption 2.2.1,

$$n^{1-\kappa}\int_{[0,\tau]} I[H(s)=0]\Lambda_X(ds) \xrightarrow{P} 0; \tag{2.29}$$

Proof. According to Condition 2.1.6 and Condition 2.1.8 we know that $\mathbf{P}(C \geq s > Y)\cdot S_X(s-) > 0$ with $s \in (0,\tau)$. Since $EI[H(s)=0] = (1-\mathbf{P}(C \geq s > Y)S_X(s-))^n$, if we also have $\mathbf{P}(C \geq s > Y)\cdot S_X(s-) > 0$ at $s=0$ and $s=\tau$, then the lemma is obviously true. Thus to prove the lemma, we only need to consider the proof of part III and to show that (2.29) holds when $\mathbf{P}(C \geq s > Y)\cdot S_X(s-) = 0$ at $s=0$ and $s=\tau$, and $\mathbf{P}(C \geq s > Y)\cdot S_X(s-)$ is continuous at 0 and τ. This also implies that the function $\mathbf{P}(C \geq s > Y)\cdot S_X(s-)$ is increasing at a small neighbourhood of $[0,\varepsilon)$ and decreasing at small neighbourhood of $(\tau_0,\tau]$. We can also find such ε and τ_0 to guarantee that $\inf_{s\in[\varepsilon,\tau_0]}\mathbf{P}(C \geq s > Y)\cdot S_X(s-)$ is no less than $\mathbf{P}(C \geq \varepsilon > Y)\cdot S_X(\varepsilon-)$ and $\mathbf{P}(C \geq \tau_0 > Y)\cdot S_X(\tau_0-)$.

Assumption 2.2.1 implies that $\Lambda_X(\tau) < \infty$, Λ_X is continuous at 0 and τ and $\int_{[0,\tau]}[\mathbf{P}(C \geq s > Y)\cdot S_X(s-)]^{-1}d\Lambda_X(s) < \infty$. Therefore, we can find sequences $\varepsilon_n \leq \varepsilon$ and $\tau_n \geq \tau_0$, such that $\varepsilon_n \downarrow 0$ and $\tau_n \uparrow \tau$ as $n\to\infty$ and $\Lambda_X(\varepsilon_n) = 1/(n^{1-\kappa}\log n)$, $\Lambda_X(\tau)-\Lambda_X(\tau_n) = 1/(n^{1-\kappa}\log n)$.

Because $\int_{[0,\tau]}[\mathbf{P}(C \geq s > Y)\cdot S_X(s-)]^{-1}d\Lambda_X(s) \leq \rho < \infty$, for some positive number ρ, we know that

$$\{[\mathbf{P}(C \geq \varepsilon_n > Y)\cdot S_X(\varepsilon_n-)]\}^{-1}\cdot\Lambda_X(\varepsilon_n) \leq \int_{[0,\varepsilon_n]}[\mathbf{P}(C \geq s > Y)\cdot S_X(s-)]^{-1}d\Lambda_X(s) \leq \rho < \infty$$

and further $\mathbf{P}(C \geq \varepsilon_n > Y)\cdot S_X(\varepsilon_n-) \geq \rho^{-1}\Lambda_X(\varepsilon_n) = \rho^{-1}/(n^{1-\kappa}\log n)$. Similarly we have $\mathbf{P}(C \geq \tau_n > Y)\cdot S_X(\tau_n-)] \geq \rho^{-1}/(n^{1-\kappa}\log n)$.

On the other hand, $\mathbf{P}(C \geq s > Y)\cdot S_X(s-) = EH(s)$ and

$$\begin{aligned}
&E\left|n^{1-\kappa}\int_{[0,\tau]} I[H(s)=0]\Lambda_X(ds)\right| = n^{1-\kappa}\int_{[0,\tau]}[1-EH(s)]^n\Lambda_X(ds)\\
&= n^{1-\kappa}\int_{[0,\varepsilon_n]}[1-EH(s)]^n\Lambda_X(ds) + n^{1-\kappa}\int_{[\tau_n,\tau]}[1-EH(s)]^n\Lambda_X(ds) + n^{1-\kappa}\int_{[\varepsilon_n,\tau_n]}[1-EH(s)]^n\Lambda_X(ds)\\
&\leq n^{1-\kappa}\Lambda_X(\varepsilon_n) + n^{1-\kappa}(\Lambda_X(\tau)-\Lambda_X(\tau_n)) + n^{1-\kappa}[1-\rho^{-1}/(n^{1-\kappa}\log n)]^n\Lambda_X(\tau)\\
&= \frac{2}{\log n} + O\left(n^{1-\kappa}\exp(-\rho n^{\kappa}/\log n)\right) \to 0
\end{aligned}$$

which implies (2.29). The lemma is proved. ■

Theorem 2.2. *Under Condition 2.1.6 and Condition 2.1.8, we have the following statements:*

I *If* $\mathbf{P}(C \geq \tau > Y)S_X(\tau-) > 0$, *under parts [1.] and [2.] of Assumption 2.2.1, we have for* $t \in [0,\tau]$,

$$\sqrt{n}(\hat{\Lambda}_X(t)-\Lambda_X(t)) \Rightarrow Z(t) \tag{2.30}$$

where $Z(t)$ *is a zero-mean Gaussian process with independent increments and variance function*

$$\sigma_\Lambda^2(t) = \int_0^t [\mathbf{P}(C \geq s > Y)\cdot S_X(s-)]^{-1}(1-\Delta\Lambda_X(s))d\Lambda_X(s). \tag{2.31}$$

II *If* $\mathbf{P}(C \geq \tau > Y)S_X(\tau-) = 0$, *under parts [1.], [2.] and [3.] of Assumption 2.2.1, the above result is still true.*

Proof. The proof is based on Theorem 2.1. We can write $\sqrt{n}(\hat{\Lambda}_X(t) - \Lambda_X(t)) = \int_{[0,t]} \mathbf{H}(s)dM(s) + o_p(1)$, where $\mathbf{H}(s) = \sqrt{n}/H(s) \cdot I[H(s) > 0]$.

Consider part **I** of the theorem. From the definition of $H(s)$, we know that $\sup_{s\in[0,\tau]} |n^{-1}H(s) - \mathbf{P}(C \geq s > Y) \cdot S_X(s-)| \to 0$ in probability. Therefore we have that $\mathbf{H}(s)^2 H(s) = n/H(s) \cdot I[H(s) > 0]$ satisfies, for any $0 < \varepsilon < t < \tau$,

$$\sup_{s\in[\varepsilon,t]} |\mathbf{H}(s)^2 H(s) - h(s)| \to 0$$

where $h(s) := [\mathbf{P}(C \geq s > Y) \cdot S_X(s-)]^{-1}$. For any $s \in [\varepsilon, \tau]$, $h(s)$ is bounded. Therefore part (b) of Condition 2.1.9 is satisfied. Note that if $\mathbf{P}(C \geq \tau > Y)S_X(\tau-) > 0$ the above convergence is true for $t = \tau = \bar{t}$ and $\bar{t} \in \mathscr{L}$.

In addition, part [1.] of Assumption 2.2.1 implies that (a) and (c) of Condition 2.1.9 is satisfied.

Part [2.] of Assumption 2.2.1 implies that Λ_X is continuous at 0 and therefore we have

$$\begin{aligned} & E\int_{[0,\varepsilon]} \mathbf{H}(s)^2 H(s) d\Lambda_X(s) \\ &= E\int_{[0,\varepsilon]} \frac{n}{H(s)} I[H(s) > 0] d\Lambda_X(s) \\ &\leq E\int_{[0,\varepsilon]} 2\frac{n+1}{H(s)+1} I[H(s) > 0] d\Lambda_X(s) \\ &\leq 2\int_{[0,\varepsilon]} \frac{1}{EH(s)} d\Lambda_X(s) = 2\int_{[0,\varepsilon]} \frac{1}{\mathbf{P}(C \geq s > Y) \cdot S_X(s-)} d\Lambda_X(s) < \infty \end{aligned} \tag{2.32}$$

where the second inequality uses the result of Lemma 2.3, by treating $H(s)$ as a Binomial variable of $B(n, EH(s))$. This result further implies that (2.32) converges to 0 as $\varepsilon \to 0$. Therefore part (d) of Condition 2.1.9 is satisfied. Therefore the theorem is proved for part **I**, by using Theorem 2.1.

If $\mathbf{P}(C \geq \tau > Y)S_X(\tau-) = 0$, then $\tau = \bar{t} \notin \mathscr{L}$. With parts [1.] and [3.] of Assumption 2.2.1 we can show, similarly as above, that part (e) of Condition 2.1.9 is satisfied. Therefore the theorem is proved for part (**II**), by using Theorem 2.1.

■

Lemma 2.3. *For a Binomial random variable* $\zeta \sim B(n, p)$, *then*

$$E\frac{n+1}{\zeta+1} \leq \frac{1}{p}.$$

Proof.

$$\begin{aligned} E\frac{n+1}{\zeta+1} &= \sum_{k=0}^{n} \frac{n+1}{k+1} \binom{n}{k} p^k (1-p)^{n-k} \\ &= \frac{1}{p} \sum_{k=0}^{n} \binom{n+1}{k+1} p^{k+1} (1-p)^{n-k} \leq \frac{1}{p}. \end{aligned}$$

■

A consistent estimator for $\sigma_\Lambda^2(t)$ is given by

$$n\int_0^t \frac{I[H(s) > 0]}{H(s)^2} \left(1 - \frac{\Delta N(s)}{H(s)}\right) N(ds) \tag{2.33}$$

where $\Delta N(s) = N(s) - N(s-)$.

Remark 2.1. Note that part [2.] in Assumption 2.2.1 means that that S_X is continuous at 0. We need this assumption because if F_X (or Λ_X) has an atom mass at 0, (2.32) will not converges to 0 (due to $\Delta\Lambda_X(0) > 0$ and $\mathbf{P}(C \geq 0 > Y) = 0$). This corresponds to that with our truncation condition $X > Y$, we simply cannot observe $X = 0$ and $\Delta\Lambda_X(0)$ cannot be identified. If F_X is continuous at 0, then part [1.] in Assumption 2.2.1 is in fact implied by $\int_{(0,\varepsilon)}[\mathbf{P}(C \geq s > Y-)]^{-1}dF_X(s) < \infty$, which tells us the requirement at the left tail.

If F_X does have an atom mass at time 0, the convergence of the above theorem is still true in the sense of $\sqrt{n}\int_{(\varepsilon,t)}(d\hat{\Lambda}_X(s) - \Lambda_X(s))$, for any $\varepsilon > 0$. The methodology is equivalent to dealing with the conditional distribution $\tilde{F_X}(t) = \mathbf{P}(0 < X \leq t | X > 0) = 1 + \frac{S(t)}{\Delta S(0)}$. This, however, does not affect the estimation of Λ_X since a simple calculation implies $d\Lambda_X = d\tilde{\Lambda_X}$, but $S_X(0)$ cannot be identified.

Remark 2.2. If $\int_{(t,\tau)}[\mathbf{P}(C \geq s > Y)S_X(s-)]^{-1}d\Lambda_X(s) = \infty$ (item [1.] of Assumption 2.2.1 does not hold), then the above theorem does not hold. Such scenarios correspond to heavy censoring at the right tails of F_X, therefore in practice the estimation is usually very poor at the right tails. A typical example is that $\mathbf{P}(C \geq \tau > Y) = 0$ and it is continuous at τ, but $S_X(t)$ has a jump at τ.

2.2.2 Estimation for the survival function

Lemma 2.4. *If $S_X(t) > 0$,*

$$\frac{\hat{S}_X(t)}{S_X(t)} = 1 - \int_0^t \frac{\hat{S}_X(s-)}{S_X(s)} \left\{ \frac{N(ds)}{H(s)} - \Lambda_X(ds) \right\}. \tag{2.34}$$

The bias of $\hat{S}_X(t)$ is then given by

$$\begin{aligned}
\hat{S}_X(t) - S_X(t) &= -S(t) \int_0^t \frac{\hat{S}_X(s-)}{S_X(s)} \left\{ \frac{N(ds)}{H(s)} - I[H(s) > 0]\Lambda_X(ds) - I[H(s) = 0]\Lambda_X(ds) \right\} \\
&= -S(t) \int_0^t \frac{\hat{S}_X(s-)}{S_X(s)} \frac{I[H(s) > 0]}{H(s)} M(ds) + B(t)
\end{aligned} \tag{2.35}$$

where

$$B(t) = S(t) \int_0^t \frac{\hat{S}_X(s-)}{S_X(s)} I[H(s) = 0]\Lambda_X(ds). \tag{2.36}$$

The variance of $\hat{S}_X(t)$ is given by

$$\begin{aligned}
Var(\hat{S}_X(t)) &= E\left\{\hat{S}_X(t) - S_X(t) - B(t)\right\}^2 \\
&\approx E\left\{ S(t) \int_0^t \frac{\hat{S}_X(s-)}{S_X(s)} \frac{I[H(s) > 0]}{H(s)} M(ds) \right\}^2 \\
&= S(t)^2 \int_0^t E\left\{ \frac{\hat{S}_X(s-)}{S_X(s)} \right\}^2 \frac{I[H(s) > 0]}{H(s)} (1 - \Lambda_X(ds))\Lambda_X(ds)
\end{aligned} \tag{2.37}$$

whose consistent estimator is

$$n\hat{S}(t)^2 \int_0^t \frac{dN(s)}{(H(s) - \Delta N(s))H(s)}. \tag{2.38}$$

Similarly as the results in the previous section, we have the following theorem.

Theorem 2.3. *Under Condition 2.1.6 and Condition 2.1.8, we have the following statements:*

I *If* $\mathbf{P}(C \geq \tau > Y)S_X(\tau-) > 0$, *under parts [1.] and [2.] of Assumption 2.2.1, we have for* $t \in [0, \tau]$,

$$\sqrt{n}(\hat{S}_X(t) - S_X(t)) \Rightarrow Z(t) \tag{2.39}$$

where $Z(t)$ *is a zero-mean Gaussian process with independent increments and variance function*

$$\sigma_S^2(t) = S_X(t)^2 \int_0^t \left[\frac{S_X(s-)}{S_X(s)}\right]^2 \frac{1}{\mathbf{P}(C \geq s > Y) \cdot S_X(s-)}(1 - \Delta\Lambda_X(s))d\Lambda_X(s). \tag{2.40}$$

II *If* $\mathbf{P}(C \geq \tau > Y)S_X(\tau-) = 0$, *under parts [1.], [2.] and [3.] of Assumption 2.2.1, the above result is still true.*

2.2.3 Discussion on the risk set under truncation

Under the censoring-only framework, the risk set at time $t = \tilde{X}_i$, denoted as $\mathscr{R}_{\tilde{X}_i}$, is the set of subjects which survive longer than $\tilde{X}_i$. Therefore as long as $\tilde{X}_i$ is not the maximum observation time, the risk set $\mathscr{R}_{\tilde{X}_i}$ should have more than one element. However, under truncation framework, the risk set $\mathscr{R}_{\tilde{X}_i}$ is the set of subjects $\{j : \text{such that } \tilde{X}_j \geq \tilde{X}_i > Y_j\}$ and it may have just one element even if $\tilde{X}_i$ is not the maximum point. Actually, for any observation $\tilde{X}_i$, its risk set may just have one element if the sample size is not large enough. This will cause some problem in practice. For example, the survival function estimate (see (2.8) and (2.10))

$$\hat{S}_X(t) = \prod_{s \in (0,t]} \left[1 - \frac{\Delta N(s)}{H(s)}\right] \tag{2.41}$$

may be equal to 0 at some very small value t_1, if there is an $\tilde{X}_i < t_1$ such that $H(\tilde{X}_i) = 1$ and $\Delta N(\tilde{X}_i) = 1$. Then for all $t > t_1$ we have $\hat{S}(t) = 0$. This could cause severely biased estimate for small samples. In addition, the consistent estimator (2.38) for $\text{Var}(\hat{S}_X(t))$ is infinity, since $H(s) - \Delta N(s) = 1 - 1 = 0$.

Although, for large sample sizes $H(s)$ should be much larger than $\Delta N(s)$ and such problem will not occur, we still need a way to solve this problem for small sample sizes. In fact this is not difficult, if we consider the amended product limit estimator, at time t smaller than the maximum event time. For simplicity, we assume that $\Delta N(t) \leq 1$ (no ties). Then we can estimate S_X using

$$\hat{S}_X(t) = \prod_{s \in (0,t]} \left[1 - \frac{I[H(s) > 1]\Delta N(s)}{H(s)}\right] \tag{2.42}$$

which is consistent and asymptotically normal. Its variance $\text{Var}(\hat{S}_X(t))$ can be estimated by

$$n\hat{S}(t)^2 \int_0^t \frac{I[H(s) > 1]dN(s)}{(H(s) - I[H(s) > 1]\Delta N(s))H(s)}. \tag{2.43}$$

The large sample properties for these revised estimates can be obtained similarly.

2.3 Linear Rank Statistics for umbrella alternative hypothesis

2.3.1 Preliminaries

The (weighted) log-rank test is the most commonly used statistical test for comparing the survival functions of two or more treatment groups. We here consider a very general hypothesis test for the survival functions. For example, in the dose-response research in clinical trials or animal experiments, the treatment effects may become better as the dose level increases. However, if the dose level is too high there could be an overdose effect. Consider a dose-response research with K treatment dose levels and the dose levels increase from group 1 to group K. Denote the survival function of group k as S_k. We discuss how to test the following hypotheses under censoring and truncation,

$$\begin{aligned} \mathrm{H}_0 &: S_1 = \cdots = S_K \\ \mathrm{H}_1 &: S_1 \le \cdots \le S_q \ge \cdots \ge S_K \end{aligned} \tag{2.44}$$

with at least one strict inequality in the above equation. Treatment group q corresponds the best dose level. Such a hypothesis test is called *umbrella* test (Chen and Wolfe, 2000) where q is called the peak of the umbrella.

Suppose that the censored and truncated data set of the uth treatment group is $(\tilde{X}_{1,u}, Y_{1,u}, \delta_{1,u}), \cdots, (\tilde{X}_{n_u,u}, Y_{n_u,u}, \delta_{n_u,u})$. A typical unbiased subject from the population of group u is denoted as $(\tilde{X}_{\cdot,u}, Y_{\cdot,u}, \delta_{\cdot,u})$. Denote $N_{i,u}(t) = I[\tilde{X}_{i,u} \le t, \delta_{i,u} = 1]$ as the indicator counting process for the ith subject in group u. Denote

$$\begin{aligned} N_{\cdot,u}(t) &= \sum_{i=1}^{n_u} N_{i,u}(t) \\ N_{\cdot,j}^{(1)}(t) &= \sum_{u=1}^{j} \sum_{i=1}^{n_u} N_{i,u}(t) \\ N_{\cdot,j}^{(2)}(t) &= \sum_{u=j}^{K} \sum_{i=1}^{n_u} N_{i,u}(t). \end{aligned}$$

We use similar notations for the indicator risk processes, i.e. denote $H_{i,u}(t) = I[\tilde{X}_{i,u} \ge t > Y_{i,u}]$ as the indicator risk process for the ith subject in group u. Denote $H_{\cdot,u}(t) = \sum_{i=1}^{n_u} H_{i,u}(t)$, $H_{\cdot,j}^{(1)}(t) = \sum_{u=1}^{j} \sum_{i=1}^{n_u} H_{i,u}(t)$ and $H_{\cdot,j}^{(2)}(t) = \sum_{u=j}^{K} \sum_{i=1}^{n_u} H_{i,u}(t)$. We also need to define the risk indicator process without considering the truncation variable, i.e. $\tilde{H}_{i,u}(t) = I[\tilde{X}_{i,u} \ge t]$, $\tilde{H}_{\cdot,u}(t) = \sum_{i=1}^{n_u} \tilde{H}_{i,u}(t)$, $\tilde{H}_{\cdot,j}^{(1)}(t) = \sum_{u=1}^{j} \sum_{i=1}^{n_u} \tilde{H}_{i,u}(t)$ and $\tilde{H}_{\cdot,j}^{(2)}(t) = \sum_{u=j}^{K} \sum_{i=1}^{n_u} \tilde{H}_{i,u}(t)$.

Define the following martingales

$$M_{\cdot,u}(ds) = N_{\cdot,u}(ds) - H_{\cdot,u}(s)\Lambda_u(ds)$$

where Λ_u is the cumulative hazard function corresponding to S_u, and $M_{\cdot,j}^{(1)} = \sum_{u=1}^{j} M_{\cdot,u}$, $M_{\cdot,j}^{(2)} = \sum_{u=j}^{K} M_{\cdot u}$. Let $n = \sum_{u=1}^{K} n_u$ and denote $\mathscr{L}_u = \{t : \lim_{n\to\infty} n^{-1} H_{\cdot,u}(t) > 0\}$, $\mathscr{L}_u^{(1)} = \{t : n^{-1} \lim_{n\to\infty} H_{\cdot,u}^{(1)}(t) > 0\}$, $\mathscr{L}_u^{(2)} = \{t : \lim_{n\to\infty} n^{-1} H_{\cdot,u}^{(2)}(t) > 0\}$ and $\mathscr{L} = \{t : \prod_{u=1}^{K} \lim_{n\to\infty} [n^{-1} H_{\cdot,u}(t)] > 0\}$. Suppose that $\mathscr{L}_1 = \cdots = \mathscr{L}_K = \mathscr{L} = \mathscr{L}_u^{(1)} = \mathscr{L}_u^{(2)}$. Let $\bar{t} = \sup \mathscr{L}$ and $\underline{t} = \inf \mathscr{L}$. With similar assumptions as in early sections, we may also write $\bar{t} = \tau$ and $\underline{t} = 0$.

Consider K bounded predictable process $\mathbf{H}_u(s), u = 1, \cdots, K$.

Condition 2.3.1 *(a)For any $t \in \mathscr{L}$, $\Lambda_u(t) < \infty$, for each $u = 1, \cdots, K$*
(b)There exists non-negative and left-continuous functions with right-hand limits, $h_u(s)$, such that

$$\sup_{\varepsilon \le s \le t} \left| [\mathbf{H}_u(s)]^2 H_u(s) - h_u(s) \right| \xrightarrow{P} 0, \quad n \to \infty,$$

and $\boldsymbol{h}_u(s)$ is bounded in $[\varepsilon, t]$, whenever $\varepsilon, t \in \mathscr{L}$, and $\boldsymbol{h}_u(s) = 0, \forall s \notin \mathscr{L}$;
(c) For each u, $\int_{\mathscr{L}} \boldsymbol{h}_u(s)(1-\Delta\Lambda_u(s))d\Lambda_u(s) < \infty$, when $\bar{t} \notin \mathscr{L}$;
(d) For each u, $\lim_{\varepsilon\downarrow 0} \limsup_{n\to\infty} P\left\{\int_{[0,\varepsilon]} [\mathbf{H}_u(s)]^2 H_u(s) d\Lambda_u(s) > \xi\right\} = 0, \forall \xi > 0$;
(e) For each u, $\lim_{t\uparrow\tau} \limsup_{n\to\infty} P\left\{\int_{[t,\tau]} [\mathbf{H}_u(s)]^2 H_u(s) d\Lambda_u(s) > \xi\right\} = 0, \forall \xi > 0$, when $\bar{t} \notin \mathscr{L}$.

Then we have the next theorem,

Theorem 2.4. *Let $Z_1(t), \cdots, Z_K(t)$ be independent zero-mean Gaussian processes with independent increments and continuous sample paths and variance functions $\int_{[0,t]} \boldsymbol{h}_u(s)(1-\Delta\Lambda_u(s))d\Lambda_u(s)$ and denote*

$$\boldsymbol{R}_u(t) = \int_{[0,t]} \mathbf{H}_u(s) dM_{\cdot,u}(s)$$

Under (a) and (b) of Condition 2.3.1 we have that

$$\boldsymbol{R}_u \Rightarrow Z_u,$$

on $D[\varepsilon, t]$ for any $0 < \varepsilon < t < \bar{t}$.
Under (a), (b), (c) and (d) of Condition 2.3.1 we have that

$$\boldsymbol{R}_u \Rightarrow Z_u,$$

on $D[0,t]$, for any $0 < t < \bar{t}$.
Under all items of Condition 2.3.1 the convergence is on $D[0,\tau]$, as $n \to \infty$.

2.3.2 *q* is known

For the umbrella test in (2.44), if q is known, we can use the following weighted log-rank statistic,

$$\begin{aligned}
Q_q &= \sum_{j=2}^{q} A_j^{(1)} + \sum_{j=q}^{K-1} A_j^{(2)}, \\
A_j^{(1)} &= \int_0^\tau W_j^{(1)}(s) \frac{\tilde{H}_{\cdot,j}(s)\tilde{H}_{\cdot,j-1}^{(1)}(s)}{\tilde{H}_{\cdot,j}^{(1)}(s)} \left\{ \frac{dN_{\cdot,j-1}^{(1)}(s)}{H_{\cdot,j-1}^{(1)}(s)} - \frac{dN_{\cdot,j}(s)}{H_{\cdot,j}(s)} \right\}, \quad j = 2, \cdots, q, \\
A_j^{(2)} &= \int_0^\tau W_j^{(2)}(s) \frac{\tilde{H}_{\cdot,j}(s)\tilde{H}_{\cdot,j+1}^{(2)}(s)}{\tilde{H}_{\cdot,j}^{(2)}(s)} \left\{ \frac{dN_{\cdot,j+1}^{(2)}(s)}{H_{\cdot,j+1}^{(2)}(s)} - \frac{dN_{\cdot,j}(s)}{H_{\cdot,j}(s)} \right\}, \quad j = q, \cdots, K-1. \qquad (2.45)
\end{aligned}$$

As Chen and Wolfe (2000) suggested, we can choose the weights $W_j^{(1)}(s) = \left[\hat{S}_j^{(1)}\right]^\rho$ and $W_j^{(2)}(s) = \left[\hat{S}_j^{(2)}\right]^\rho$, $\rho > 0$, where the product-limit estimates $\hat{S}_j^{(\cdot)}$ are given by,

$$\begin{aligned}
\hat{S}_j^{(1)}(t) &= \prod_{s\le t} \left\{ 1 - \frac{dN_{\cdot,j}^{(1)}(s)}{H_{\cdot,j}^{(1)}(s)} \right\} \\
\hat{S}_j^{(2)}(t) &= \prod_{s\le t} \left\{ 1 - \frac{dN_{\cdot,j}^{(2)}(s)}{H_{\cdot,j}^{(2)}(s)} \right\}.
\end{aligned}$$

Under the null hypothesis, $\hat{S}_j^{(1)}(t)$ is the product-limit estimate based on the first j groups and $\hat{S}_j^{(2)}(t)$ is the product-limit estimate based on the last $K-j+1$ groups. The statistic Q_q degenerates to the two-sample test if $K=2$ or to the K-group ordered test if $q=1$ or $q=K$ (Fleming et al., 1987).

We can also write Q_q in (2.45) as

$$Q_q = \sum_{u=1}^{K} \int_0^{\tau} \mathbf{H}_u^{(q)}(s)dM_u(s) + \sum_{u=1}^{K} \int_0^{\tau} \mathbf{K}_u^{(q)}(s)d\Lambda_u(s) \tag{2.46}$$

where

$$\mathbf{H}_u^{(q)} = \begin{cases} \left[\sum_{j=u+1}^{q} W_j^{(1)} \frac{\tilde{H}_{\cdot,j}\tilde{H}^{(1)}_{\cdot,j-1}}{\tilde{H}^{(1)}_{\cdot,j}} \frac{I(\cdot;\mathscr{L}^{(1)}_{j-1})}{H^{(1)}_{\cdot,j-1}}\right] - W_u^{(1)} \frac{\tilde{H}_{\cdot,u}\tilde{H}^{(1)}_{\cdot,u-1}}{\tilde{H}^{(1)}_{\cdot,u}} \frac{I(\cdot;\mathscr{L}_u)}{H_{\cdot,u}}, & 1 \le u \le q-1 \\ -W_q^{(1)} \frac{\tilde{H}_{\cdot,q}\tilde{H}^{(1)}_{\cdot,q-1}}{\tilde{H}^{(1)}_{\cdot,q}} \frac{I(\cdot;\mathscr{L}_q)}{H_{\cdot,q}} - W_q^{(2)} \frac{\tilde{H}_{\cdot,q}\tilde{H}^{(2)}_{\cdot,q+1}}{\tilde{H}^{(2)}_{\cdot,q}} \frac{I(\cdot;\mathscr{L}_q)}{H_{\cdot,q}}, & u=q \\ \left[\sum_{j=q}^{u-1} W_j^{(2)} \frac{\tilde{H}_{\cdot,j}\tilde{H}^{(2)}_{\cdot,j+1}}{\tilde{H}^{(2)}_{\cdot,j}} \frac{I(\cdot;\mathscr{L}^{(2)}_{j+1})}{H^{(2)}_{\cdot,j+1}}\right] - W_u^{(2)} \frac{\tilde{H}_{\cdot,u}\tilde{H}^{(2)}_{\cdot,u+1}}{\tilde{H}^{(2)}_{\cdot,u}} \frac{I(\cdot;\mathscr{L}_u)}{H_{\cdot,u}}, & q+1 \le u \le K, \end{cases}$$

$I(s;\mathscr{L})$ is the indicator function with value 1 if $s \in \mathscr{L}$ and with value 0 elsewhere and

$$\mathbf{K}_u^{(q)} = \mathbf{H}_u^{(q)} \cdot H_{\cdot,u} = \begin{cases} \left[\sum_{j=u+1}^{q} W_j^{(1)} \frac{\tilde{H}_{\cdot,j}\tilde{H}^{(1)}_{\cdot,j-1}}{\tilde{H}^{(1)}_{\cdot,j}} \frac{I(\cdot;\mathscr{L}^{(1)}_{j-1})}{H^{(1)}_{\cdot,j-1}}\right] H_{\cdot,u} - W_u^{(1)} \frac{\tilde{H}_{\cdot,u}\tilde{H}^{(1)}_{\cdot,u-1}}{\tilde{H}^{(1)}_{\cdot,u}} I(\cdot;\mathscr{L}_u), & 1 \le u \le q-1 \\ -W_q^{(1)} \frac{\tilde{H}_{\cdot,q}\tilde{H}^{(1)}_{\cdot,q-1}}{\tilde{H}^{(1)}_{\cdot,q}} I(\cdot;\mathscr{L}_q) - W_q^{(2)} \frac{\tilde{H}_{\cdot,q}\tilde{H}^{(2)}_{\cdot,q+1}}{\tilde{H}^{(2)}_{\cdot,q}} I(\cdot;\mathscr{L}_q), & u=q \\ \left[\sum_{j=q}^{u-1} W_j^{(2)} \frac{\tilde{H}_{\cdot,j}\tilde{H}^{(2)}_{\cdot,j+1}}{\tilde{H}^{(2)}_{\cdot,j}} \frac{I(\cdot;\mathscr{L}^{(2)}_{j+1})}{H^{(2)}_{\cdot,j+1}}\right] H_{\cdot,u} - W_u^{(2)} \frac{\tilde{H}_{\cdot,u}\tilde{H}^{(2)}_{\cdot,u+1}}{\tilde{H}^{(2)}_{\cdot,u}} I(\cdot;\mathscr{L}_u), & q+1 \le u \le K. \end{cases}$$

The above defined $\mathbf{H}_u(s)$ is a bounded predictable process, depending on the sample size n. It is easy to show that, in probability,

$$h_u^{(q)}(s) = \lim_{n\to\infty} \frac{\left[\mathbf{H}_u^{(q)}(s)\right]^2 H_{\cdot,u}(s)}{n} \tag{2.47}$$

$$k_u^{(q)}(s) = \lim_{n\to\infty} \frac{\mathbf{K}_u^{(q)}(s)}{n}. \tag{2.48}$$

The asymptotic distribution of Q_q is then given by the following theorem.

Assumption 2.3.1 *(a)For every $u=1,\cdots,K$ and $n=\sum_{u=1}^{K} n_u$, the limit $0<\lim_{n\to\infty} n_u/n<1$ exists;*
(b)For every u, $\int_{[0,\tau)} d\Lambda_u(ds)<\infty$ and $\int_{(0,\tau)} [\mathbf{P}(C_{\cdot,u} \ge s > Y_{\cdot,u}) \cdot S_u(s-)]^{-1} d\Lambda_u(s) < \infty$;
(c)For every u, item 2 of Condition 2.1.3 is satisfied;
(d)For every u, item 1 of Condition 2.1.3 is satisfied.

Theorem 2.5. *If, for every $u=1,\cdots,K$, Condition 2.1.6 and Condition 2.1.8 hold, then we have the following statements:*

I *If $\mathbf{P}(C_{\cdot,u} \ge \tau > Y_{\cdot,u}) \cdot S_u(\tau-) > 0$ and parts (a), (b) and (c) of Assumption 2.3.1 are true, then*

$$n^{-1/2} \sum_{u=1}^{K} \int_{[0,\tau]} \mathbf{H}_u^{(q)}(s)dM_u(s) \xrightarrow{d} N(0,\sigma_q^2) \tag{2.49}$$

as $n\to\infty$, where $\sigma_q^2 = \sum_{u=1}^{K} \int_{[0,\tau]} h_u^{(q)}(s)(1-\Delta\Lambda_u(s))d\Lambda_u(s)$. We also have, under the null hypothesis H_0,

$$\lim_{n\to\infty} n^{-1/2} \sum_{u=1}^{K} \int_{[0,\tau]} \mathbf{K}_u^{(q)}(s) d\Lambda_u(s) \xrightarrow{P} 0 \tag{2.50}$$

as $n \to \infty$.

II *If $\mathbf{P}(C_{\cdot,u} \geq \tau > Y_{\cdot,u}) \cdot S_u(\tau-) = 0$ and parts (a), (b), (c) and (d) of Assumption 2.3.1 are true, then the above results still hold.*

■

Proof. The proof follows from Theorem 2.4.

Under H_0, a consistent estimate for σ_q^2 is given by

$$\hat{\sigma}_q^2 = n^{-1} \sum_{u=1}^{K} \int_{[0,\tau]} \left[\mathbf{H}_u^{(q)}(s)\right]^2 H_{\cdot,u}(s) \left(1 - \frac{\Delta N_{\cdot,\cdot}(s)}{H_{\cdot,\cdot}(s)}\right) \frac{dN_{\cdot,\cdot}(s)}{H_{\cdot,\cdot}(s)} \tag{2.51}$$

where $N_{\cdot,\cdot} = \sum_{u=1}^{K} N_{\cdot,u}$ and $H_{\cdot,\cdot} = \sum_{u=1}^{K} H_{\cdot,u}$.

From Theorem 2.5, we know that under H_0, the statistic $n^{-1/2} Q_q / \sqrt{\hat{\sigma}_q^2}$ weakly converges to the standard normal distribution. Therefore, considering the hypothesis test with significance level α, the null hypothesis will be rejected if $n^{-1/2} Q_q / \sqrt{\hat{\sigma}_q^2} \geq z_{1-\alpha}$, where z_α is the $\alpha \cdot 100\%$ quantile of the standard normal distribution.

2.3.3 *q* is unknown

For the umbrella test in (2.44), if q is unknown, the hypothesis test model in (2.44) can be viewed as a combined model of K sub-models. We can choose the test statistic as

$$Q_{\max} = n^{-1/2} \max\left\{ \frac{Q_1}{\sqrt{\hat{\sigma}_1^2}}, \cdots, \frac{Q_K}{\sqrt{\hat{\sigma}_K^2}} \right\}$$

The asymptotic distribution of $Q_{\max}$ can be find out numerically based on the result

$$\left(\frac{Q_1}{\sqrt{\hat{\sigma}_1^2}}, \cdots, \frac{Q_K}{\sqrt{\hat{\sigma}_K^2}} \right) \xrightarrow{d} N(\mathbf{0}, \boldsymbol{\Sigma})$$

where the diagonal elements in $\boldsymbol{\Sigma}$ are 1s and the element in the jth row and kth column is $\sigma_{jk}/(\sigma_j \sigma_k)$ with $\sigma_{jk} = \sum_{u=1}^{K} \int_{[0,\tau]} h_u^{(jk)}(s)(1 - \Delta\Lambda_u(s)) d\Lambda_u(s)$,

$$h_u^{(jk)} = \lim_{n\to\infty} \frac{\mathbf{H}_u^{(j)} \mathbf{H}_u^{(k)} H_{\cdot,u}}{n}.$$

Under H_0, a consistent estimate for σ_{jk} is given by

$$\hat{\sigma}_{jk} = n^{-1} \sum_{u=1}^{K} \int_{[0,\tau]} \left[\mathbf{H}_u^{(j)}(s) \mathbf{H}_u^{(k)}(s)\right] H_{\cdot,u}(s) \left(1 - \frac{\Delta N_{\cdot,\cdot}(s)}{H_{\cdot,\cdot}(s)}\right) \frac{dN_{\cdot,\cdot}(s)}{H_{\cdot,\cdot}(s)} \tag{2.52}$$

Then given a significance level α, the rejection region for the test statistic $Q_{\max}$ can be calculated numerically.

2.4 Regression analysis for truncated and censored data

In the previous sections, we have studied the estimation of survival functions of homogeneous groups. For studies where there are several treatment groups, a hypothesis test method was introduced for the comparison of the survival functions in different groups. In practice, it is often more convenient to model the relation of survival times and different treatment methods (or other explanatory variables) via regression models.

A widely used approach is the proportional hazard regression model, which aims to model how hazard rate is affected by explanatory variables. It has the particular form,

$$\lambda(t|Z) = \lambda_0(t)\exp(\beta' Z)$$

where Z is a vector of explanatory variables and $\lambda_0(t)$ is a baseline hazard rate function, which is completely unspecified. A subject-based hazard is proportional to the baseline hazard, with a multiple factor $\exp(\beta' Z)$. Therefore a unit change in one explanatory variable Z implies a change in the hazard rate by a multiple factor $\exp(\beta)$, which is called the hazard ratio. The analysis of proportional hazard models focuses on the estimation of parameters β (then the estimation of hazard ratio) and the baseline hazard λ_0. Such a model is *semiparametric*, as it is neither fully parametric nor nonparametric.

In this section, we provide a brief introduction of the methodologies for proportional hazard regression models under both censoring and truncation.

2.4.1 Cox regression models

Consider the censored and truncated observations $(\tilde{X}_i, Y_i, \delta_i, Z_i), i = 1, \cdots, n$, where Z_i is the covariate vector. Note that covariates could be either time-independent or time-dependent. Here we use the time-dependent notation $\{Z_i(t), t \in [0, \tau]\}$ for the covariate vector process to develop the methodology in a general framework. In practice, the covariate process $Z_i(t)$ could be observed intermittently and with error. In that situation, we should consider a joint modelling approach for the survival events and repeated measurements (see Su and Wang, 2012), which is discussed in Chapter 5.

The covariate process $Z_i(t)$ can be continuous or, more generally it is usually right-continuous with left-hand limit. For example, if $Z_i(t)$ represents cumulative dose levels and the medicine is given to the subject once every week, then $Z_i(t)$ will be right-continuous step-wise function with left-hand limit. Define $\mathscr{G}_{i,t} = \sigma\{Z_i(s); 0 \le s \le t\}$. If we drop the subscript i, the notation $\mathscr{G}_t$ means the filtration for a typical subject in the whole population (without truncation bias). Similarly as before, the biased sample $(\tilde{X}_i, Y_i, \delta_i, \mathscr{G}_{i,t}), i = 1, \cdots, n$ can be viewed as a subsequence of all subjects in the population, $(\tilde{X}_i^*, Y_i^*, \delta_i^*, \mathscr{G}_{i,t}^*), i = 1, \cdots$. Suppose the ith subject in the data corresponds to the j_ith observation in the population.

We here assume that $X \perp (Y, C)|\mathscr{G}_\tau$. Note this independent assumption does not imply $X_i \perp (Y_i, C_i)|\mathscr{G}_{i,\tau}$ due to truncation bias. We have that

$$\begin{aligned}
&\mathbf{P}(\tilde{X}_i \in [t, t+dt), \delta_i = 1|\mathscr{G}_{i,\tau}, \tilde{X}_i \ge t > Y_i) \\
&\mathbf{P}(\tilde{X}_i \in [t, t+dt), t > Y_i, \delta_i = 1|\mathscr{G}_{i,\tau})/\mathbf{P}(\tilde{X}_i \ge t > Y_i|\mathscr{G}_{i,\tau}) \\
&= \mathbf{P}(X_{j_i}^* \in [t, t+dt), C_{j_i}^* \ge t > Y_{j_i}^*|\mathscr{G}_{j_i,\tau}^*, \tilde{X}_{j_i}^* > Y_{j_i}^*)/\mathbf{P}(\tilde{X}_{j_i}^* \ge t > Y_{j_i}^*|\mathscr{G}_{j_i,\tau}^*, \tilde{X}_{j_i}^* > Y_{j_i}^*) \\
&= \frac{\mathbf{P}(X_{j_i}^* \in [t, t+dt), C_{j_i}^* \ge t > Y_{j_i}^*|\mathscr{G}_{j_i,\tau}^*)}{\mathbf{P}(\tilde{X}_{j_i}^* > Y_{j_i}^*|\mathscr{G}_{j_i,\tau}^*)} \Bigg/ \frac{\mathbf{P}(X_{j_i}^* \ge t, C_{j_i}^* \ge t > Y_{j_i}^*|\mathscr{G}_{j_i,\tau}^*)}{\mathbf{P}(\tilde{X}_{j_i}^* > Y_{j_i}^*|\mathscr{G}_{j_i,\tau}^*)} \\
&= \mathbf{P}(X_{j_i}^* \in [t, t+dt)|\mathscr{G}_{j_i,\tau}^*)/\mathbf{P}(X_{j_i}^* \ge t|\mathscr{G}_{j_i,\tau}^*) \qquad (2.53) \\
&= d\Lambda_X(t|\mathscr{G}_{j_i,\tau}^*) = d\Lambda_X(t|\mathscr{G}_{i,\tau})
\end{aligned}$$

which is the hazard rate function for X, depending on the covariate process. The proportional hazard regression model can be written as

$$d\Lambda_X(t|\mathscr{G}_{i,t}) = d\Lambda_0(t)\exp\left(\beta' \boldsymbol{Z}_i(t-)\right)$$

where Λ_0 is called the baseline hazard. The use of the left-limit $\boldsymbol{Z}(t-)$ in the above model means that the hazard rate at the time point t depends on the history of the $\boldsymbol{Z}$ up to, but not including, time t. For simplicity of notations, however, we simply use $\boldsymbol{Z}(t)$ throughout this book, rather than $\boldsymbol{Z}(t-)$.

Note that in (2.53), $\mathbf{P}(X > Y|\mathscr{G}_\tau)$ is the truncation probability which depends on the covariate process.

We denote F_0 and S_0 as the cumulative distribution function and survival function corresponding to Λ_0, respectively. For simplicity of notations, we assume that $a = \inf\{t : \mathbf{P}(C \geq t > Y|\mathscr{G}_s) > 0\}$ and $b = \sup\{t : \mathbf{P}(C \geq t > Y|\mathscr{G}_s) > 0\}$ do not depend on the covariate information $\mathscr{G}_s$ for any s. Naturally, the end of the study time point is defined as $\tau = b \wedge b_{F_0}$. Similarly as before, we also need the following conditions to guarantee that Λ_X is identifiable.

Condition 2.4.1 *The support for F_0 satisfies Condition 2.1.6 and Condition 2.1.8, i.e. the function $\mathbf{P}(Y < t \leq C|\mathscr{G}_\tau) > 0$ (as a function of t) almost surely with respect to dF_0.*

We use the same counting process notations in (2.12) and define the filtration $\mathscr{F}_t = \bigvee_{i=1}^n \mathscr{F}_{i,t}$,

$$\mathscr{F}_{i,t} = \sigma\{N_i(u), N_i^C(u), I[Y_i \leq u] : 0 \leq u \leq t]\} \bigvee \mathscr{G}_{i,t}, \; t \in [0,\tau]. \tag{2.54}$$

Using similar arguments as that in Lemma 2.1, we have

$$\begin{aligned} E\{dN_i(t)|\mathscr{F}_{i,t-}\} &= I[\tilde{X}_i \geq t > Y_i]E\{dN_i(t)|\mathscr{G}_{i,t-}, \tilde{X}_i \geq t > Y_i\} \\ &= I[\tilde{X}_i \geq t > Y_i]d\Lambda_i(t|\mathscr{G}_{i,t-}) \\ &= I[\tilde{X}_i \geq t > Y_i]\exp\left(\beta' \boldsymbol{Z}_i(t)\right) d\Lambda_0(t). \end{aligned} \tag{2.55}$$

As in early section we denote $H_i(t) = I[\tilde{X}_i \geq t > Y_i]$ and $H(t) = \sum_{i=1}^n H_i(t)$. Following Andersen et al. (1993), if we treat $dN_i(t)$ as a 0-1 random variable and assume that $\sum_{i=1}^n \Delta N_i(t) \leq 1$ (no ties for the observed survival times; if n is large this assumption means that Λ_0 is continuous), the $\mathscr{F}_t$-likelihood is proportional to

$$\prod_{t\in[0,\tau]} \left\{ \prod_{i=1}^n [d\Lambda_i(t)]^{\Delta N_i(t)} \left[1 - \sum_{i=1}^n H_i(t) d\Lambda_i(t)\right]^{1-\Delta N(t)} \right\}. \tag{2.56}$$

For continuous $\Lambda_i(t)$ and using the product integration result, the likelihood can be written as

$$\begin{aligned} &\prod_{t\in[0,\tau]} \left\{ \prod_{i=1}^n [d\Lambda_i(t)]^{\Delta N_i(t)} \right\} \cdot \exp\left[-\int_0^\tau \sum_{i=1}^n H_i(t) d\Lambda_i(t) \right] \\ = &\prod_{t\in[0,\tau]} \left\{ \prod_{i=1}^n [d\Lambda_0(t)\exp\left(\beta' \boldsymbol{Z}_i(t)\right)]^{\Delta N_i(t)} \right\} \cdot \exp\left[-\int_0^\tau \sum_{i=1}^n \exp\left(\beta' \boldsymbol{Z}_i(t)\right) H_i(t) d\Lambda_0(t) \right] \end{aligned} \tag{2.57}$$

which will be maximized with $d\Lambda_0(t)$ being as $d\hat{\Lambda}_0(t) = dN(t)/S^{(0)}(\beta,t)$,

$$S^{(0)}(\beta,t) = \sum_{i=1}^n \exp\left(\beta' \boldsymbol{Z}_i(t)\right) H_i(t). \tag{2.58}$$

Note that the above integration is written as $\int_0^\tau$ since Λ_0 is continuous (with no atom point having positive mass) and the integration is Riemann integral.

Replacing Λ_0 by $\hat{\Lambda}_0$ in (2.57), we can get the partial likelihood for β,

$$L(\beta) = \prod_{t\in[0,\tau]} \prod_{i=1}^{n} \left\{ \frac{\exp(\beta' Z_i(t))}{S^{(0)}(\beta,t)} \right\}^{\Delta N_i(t)}. \tag{2.59}$$

This is the Cox partial likelihood for truncated and censored data. Note that even if there are ties (F_0 can assign positive weights for some atom points), equation (2.59) can still be used as the partial likelihood and the estimate will be good if the number of ties $\Delta N(t)$ is small relative to the size of risk set $H(t)$.

To find the estimate of β it is more convenient to maximize the log-partial likelihood function

$$\log L(\beta) = \sum_{i=1}^{n} \int_{t\in[0,\tau]} \left\{\beta' Z_i(t) - \log S^{(0)}(\beta,t)\right\} dN_i(t) \tag{2.60}$$

or equivalently to solve the equation that the following score function is equal to zero,

$$U(\beta,\tau) = \frac{\partial \log L(\beta)}{\partial \beta} = \sum_{i=1}^{n} \int_{t\in[0,\tau]} \left\{ Z_i(t) - \frac{S^{(1)}(\beta,t)}{S^{(0)}(\beta,t)} \right\} dN_i(t) = 0 \tag{2.61}$$

where

$$S^{(1)}(\beta,t) = \frac{\partial S^{(0)}(\beta,t)}{\partial \beta} = \sum_{i=1}^{n} H_i(t) Z_i(t) \exp(\beta' Z_i(t)). \tag{2.62}$$

If we define the matrix,

$$S^{(2)}(\beta,t) = \frac{\partial S^{(1)}(\beta,t)}{\partial \beta} = \sum_{i=1}^{n} Z_i(t)^{\otimes 2} H_i(t) \exp(\beta' Z_i(t)), \tag{2.63}$$

we can show that,

$$\mathscr{I}(\beta,t) = -\frac{\partial U(\beta,t)}{\partial \beta} = \sum_{i=1}^{n} \int_{x\in[0,t]} \left\{ \frac{S^{(2)}(\beta,x)}{S^{(0)}(\beta,x)} - \frac{S^{(1)}(\beta,x)^{\otimes 2}}{S^{(0)}(\beta,x)^2} \right\} dN_i(x). \tag{2.64}$$

If $\mathscr{I}(\beta,\tau)$ given in (2.70) is a positive definite matrix, then the solution $\hat{\beta}$ for the above estimating equations (2.61) is indeed the maximum likelihood estimate .

An estimate for the baseline hazard function is given by

$$d\hat{\Lambda}_0(t) = \frac{dN(t)}{\sum_{i=1}^{n} H_i(s) \exp(\hat{\beta}' Z_i(t))} \tag{2.65}$$

Denote the true parameter values as β_0 and $\Lambda_{0,0}$. Now we demonstrate the consistency of the above estimates.

Condition 2.4.2 *(a)Finite baseline hazard,* $\int_0^\tau d\Lambda_{0,0}(t) < \infty$ *and* $\int_0^\tau s^{(0)}(\beta_0,t)^{-1} d\Lambda_{0,0}(t) < \infty$*;*
(b)There exists a neighbourhood $\mathscr{B}$ *of* β_0 *and the, respectively, scalar, vector and matrix functions* $S^{(j)}, j = 0,1,2$ *defined in (2.58), (2.62) and (2.63), are such that*

$$\sup_{t\in[0,\tau],\beta\in\mathscr{B}} |S^{(j)}(\beta,t) - s^{(j)}(\beta,t)| \to 0,$$

in probability as $n \to \infty$*. The limiting functions* $s^{(j)}(\beta,t)$ *are continuous with respect to* β *for each value of* t*.*
(c)The covariate $\sup_{1\le i\le n, 0\le t\le \tau} |Z_i(t)|$ *is bounded in probability.*
(d)Condition 2.4.1 holds.

(e) The matrix

$$v(\beta_0,\tau)=\int_0^{\tau}\left\{\frac{s^{(2)}(\beta_0,t)}{s^{(0)}(\beta_0,t)}-\frac{s^{(1)}(\beta_0,t)^{\otimes 2}}{s^{(0)}(\beta_0,t)^2}\right\}s^{(0)}(\beta_0,t)d\Lambda_{0,0}(t) \tag{2.66}$$

is positive definite.

We can show that

Theorem 2.6. *As $n\to\infty$, $\hat{\beta}\to\beta_0$ and $\hat{\Lambda}_0(t)\to\Lambda_{0,0}(t)$ in probability under Condition 2.4.2.*

Now we establish the asymptotic normality of the estimators. If we write the score statistic $\boldsymbol{U}$ as

$$0=\boldsymbol{U}(\hat{\beta},\tau)=\boldsymbol{U}(\beta_0,\tau)-\mathscr{I}(\beta^+,\tau)(\hat{\beta}-\beta_0)$$

where β^+ is on a line segment between $\hat{\beta}$ and β_0. Therefore we have

$$\sqrt{n}(\hat{\beta}-\beta_0)=\left(n^{-1}\mathscr{I}(\beta^+,\tau)\right)^{-1}n^{-1/2}U(\beta_0,\tau) \tag{2.67}$$

Therefore to establish the asymptotic normality of $\hat{\beta}$, it is sufficient to establish the weak convergence property for $n^{-1/2}U(\beta_0,t)$.

If we define the process $M_i(t)$ as the integration of

$$dM_i(t)=dN_i(t)-H_i(t)\exp(\beta_0'Z_i(t))d\Lambda_{0,0}(t) \tag{2.68}$$

then the score function process, evaluated at the true parameters, can be written as

$$U(\beta_0,t)=\sum_{i=1}^{n}\int_{x\in[0,t]}\left\{Z_i(x)-\frac{S^{(1)}(\beta_0,x)}{S^{(0)}(\beta_0,x)}\right\}dM_i(x) \tag{2.69}$$

since

$$\begin{aligned}&\sum_{i=1}^{n}\left\{Z_i(x)-\frac{S^{(1)}(\beta,x)}{S^{(0)}(\beta,x)}\right\}H_i(x)\exp(\beta'Z_i(x))\\&=\sum_{i=1}^{n}Z_i(x)H_i(x)\exp(\beta'Z_i(x))-\sum_{i=1}^{n}\frac{S^{(1)}(\beta,x)}{S^{(0)}(\beta,x)}H_i(x)\exp(\beta'Z_i(x))\\&=S^{(1)}(\beta,x)-S^{(1)}(\beta,x)=0.\end{aligned}$$

The result of (2.55) tells us that $M_i(t)$ is a martingale. Under Condition 2.4.2, we know the following.

Theorem 2.7. *The process $n^{-1/2}U(\beta_0,t)$ is a martingale with variance function,*

$$V(\beta_0,t)=n^{-1}\sum_{i=1}^{n}\int_{x\in[0,t]}\left\{S^{(2)}(\beta_0,x)-\frac{S^{(1)}(\beta_0,x)^{\otimes 2}}{S^{(0)}(\beta_0,x)}\right\}d\Lambda_{0,0}(x). \tag{2.70}$$

As $n\to\infty$,

$$n^{-1/2}U(\beta_0,t)\Rightarrow W(t),\quad on\ D[0,\tau]$$

where $W(t)$ is a zero-mean Gaussian process with variance function $v(\beta_0,t)$ given in (2.66)

A consistent estimate for $v(\beta,t)$ is

$$\mathscr{I}(\hat{\beta},t) = n^{-1}\sum_{i=1}^{n}\int_{x\in[0,t]}\left\{\frac{S^{(2)}(\hat{\beta},x)}{S^{(0)}(\hat{\beta},x)} - \frac{S^{(1)}(\hat{\beta},x)^{\otimes 2}}{S^{(0)}(\hat{\beta},x)^2}\right\}dN_i(x). \tag{2.71}$$

■

Then from (2.67) and $\lim_{n\to\infty} n^{-1}\mathscr{I}(\beta^+,t) = \lim_{n\to\infty} n^{-1}\mathscr{I}(\hat{\beta},t) = \boldsymbol{v}(\beta_0,\tau)$, we know that $\sqrt{n}(\hat{\beta}-\beta_0)$ is asymptotically normal with mean $\mathbf{0}$ and asymptotic variance $(\boldsymbol{v}(\beta_0,\tau))^{-1}$.

Following Fleming and Harrington (1991), we have that $\sqrt{n}(\hat{\Lambda}_0(t)-\Lambda_{0,0}(t))$ converges on $D[0,\tau]$ to a zero-mean Gaussian process and variance function

$$\int_0^t \frac{d\Lambda_{0,0}(x)}{s^{(0)}(\beta_0,x)} + \boldsymbol{q}(\beta_0,t)\boldsymbol{v}(\beta_0,t)^{-1}\boldsymbol{q}(\beta_0,t)$$

where

$$\boldsymbol{q}(\beta_0,t) = \int_0^t \frac{\boldsymbol{s}^{(1)}(\beta_0,x)}{s^{(0)}(\beta_0,x)} d\Lambda_{0,0}(x)$$

An estimate of $Var(\sqrt{n}(\hat{\Lambda}(t)-\Lambda_{0,0}(t)))$ is

$$\int_{x\in[0,t]} \frac{d\hat{\Lambda}_0(x)}{S^{(0)}(\hat{\beta},x)} + \boldsymbol{Q}(\hat{\beta},t)\mathscr{I}(\hat{\beta},t)^{-1}\boldsymbol{Q}(\hat{\beta},t)$$

where

$$\boldsymbol{Q}(\hat{\beta},t) = n^{-1}\int_{x\in[0,t]} \frac{\boldsymbol{S}^{(1)}(\hat{\beta},x)}{\hat{S}^{(0)}(\hat{\beta},x)^2} dN(x).$$

2.4.2 Model diagnostics

Assessing model adequacy using diagnostic tools plays an important role in regression analyses. In this section, we illustrate several methodologies based on the martingale residuals for accuracy assessment of the Cox model in predicting the failure rate and the validity of proportional hazards assumption.

2.4.2.1 Martingale residuals

The process $M_i(t)$ defined in (2.68) can be written as

$$M_i(t) = N_i(t) - \int_{x\in[0,t]} H_i(x)\exp\left(\beta_0' Z_i(x)\right)d\Lambda_{0,0}(x), \tag{2.72}$$

which can be interpreted as the difference between the observed number of events for the ith subject and a conditionally expected number of events over $[0,t]$. Substituting β_0 by $\hat{\beta}$ and $\Lambda_{0,0}$ by $\hat{\Lambda}_0$, we obtain a type of residual which is of the form

$$\widehat{M}_i(t) = N_i(t) - \int_{x\in[0,t]} H_i(x)\exp\left(\hat{\beta}' Z_i(x)\right)d\hat{\Lambda}_0(x), \tag{2.73}$$

and this $\widehat{M}_i(t)$ is called the *martingale residual* at time t. For any $t\in[0,\infty]$,

$$\sum_{i=1}^{n}\widehat{M}_i(t)=\sum_{i=1}^{n}\left\{N_i(t)-\int_{x\in[0,t]}H_i(x)\exp\left(\hat{\beta}'Z_i(x)\right)d\hat{\Lambda}_0(x)\right\}$$
$$=\sum_{i=1}^{n}\left\{\int_{x\in[0,t]}dN_i(x)-\int_{x\in[0,t]}H_i(x)\exp\left(\hat{\beta}'Z_i(x)\right)\left[\frac{\sum_{j=1}^{n}dN_j(x)}{\sum_{k=1}^{n}H_k(x)\exp\left(\hat{\beta}'Z_k(x)\right)}\right]\right\}$$
$$=0, \tag{2.74}$$

which indicates the martingale residuals sum to zero. And also $cov(\widehat{M}_i,\widehat{M}_j)=0=E\widehat{M}_i$ asymptotically.

For a Cox model with no time-varying covariates, the martingale residual (at time ∞) reduces to

$$\widehat{M}_i=\delta_i-\left[\hat{\Lambda}_0(\tilde{X}_i)-\hat{\Lambda}_0(Y_i)\right]\exp\left(\hat{\beta}'Z_i\right), \tag{2.75}$$

where $\delta_i=I[X_i\le C_i]$ and $\tilde{X}_i=\min(X_i,C_i)$.

2.4.2.2 Deviance residuals

In the single event setting of the Cox model (i.e. no ties at each time), since $N_i(\infty)=I[\tilde{X}_i\le\infty,\delta_i=1]\le 1$, the martingale residual $\widetilde{M}_i$ given in (2.73) has the maximum value of 1 but a minimum value $-\infty$. Therefore the martingale residuals have skewed distributions. When assessing the accuracy of prediction for each subject, transformation to a normal-shaped distribution is more desirable. One such transformation is motivated by the deviance residuals from the generalized linear model literature (McCullagh and Nelder, 1989).

The definition of deviance is

$$D=2\left\{\text{log likelihood (saturated)}-\text{log likelihood}(\hat{\beta})\right\}, \tag{2.76}$$

where a saturated model is the one that each observation is allowed its own vector of $\hat{\beta}$, denoted by b_i for the ith observed subject. Assuming non-time-varying covariates and the nuisance parameter Λ_0 is known, the deviance of a Cox model under both censoring and truncation is

$$D=2\sup_{b}\sum_{i=1}^{n}\left\{\int_{t\in[0,\tau]}\left[\log e^{b_i'Z_i}-\log e^{\hat{\beta}'Z_i}\right]dN_i(t)-\int_{t\in[0,\tau]}H_i(t)\left[\exp(b_i'Z_i)-\exp(\hat{\beta}'Z_i)\right]d\Lambda_0(t)\right\}. \tag{2.77}$$

Then $\hat{b}_i$ is the one that satisfies

$$\int_{t\in[0,\tau]}H_i(t)\exp(\hat{b}_iZ_i)d\Lambda_0(t)=\int_{t\in[0,\tau]}dN_i(t).$$

Denote the martingale residual with estimate of β and known baseline hazard Λ_0 as

$$\widetilde{M}_i=N_i(\tau)-\int_{t\in[0,\tau]}H_i(t)\exp\left(\hat{\beta}'Z_i\right)d\Lambda_0(t). \tag{2.78}$$

Then the deviance given in (2.77) can be written as

$$D=-2\sum_{i=1}^{n}\left\{\widetilde{M}_i+\log\left[\frac{\exp(\hat{\beta}'Z_i)}{\exp(b_i'Z_i)}\right]\int_{t\in[0,\tau]}dN_i(t)\right\}$$
$$=-2\sum_{i=1}^{n}\left\{\widetilde{M}_i+N_i(\tau)\log\left[\frac{N_i(\tau)-\widetilde{M}_i}{N_i(\tau)}\right]\right\}. \tag{2.79}$$

The last step above requires a factorization

$$\int_{t\in[0,\tau]} H_i(t)\exp(\hat{\beta}'Z_i)d\Lambda_0(t) = \exp(\hat{\beta}'Z_i)\int_{t\in[0,\tau]} H_i(t)d\Lambda_0(t),$$

which is not valid for time-varying covariates $Z_i(t)$.

When the nuisance parameter Λ_0 is estimated by (2.65), $\widetilde{M}_i$ in (2.79) is replaced by $\widehat{M}_i$ given in (2.75). Then the *deviance residual*, d_i, is the signed square root of (2.79),

$$d_i = \mathrm{sign}(\widehat{M}_i)\sqrt{2}\left\{\widehat{M}_i + \delta_i \log(\delta_i - \widehat{M}_i)\right\}^{1/2}. \tag{2.80}$$

The martingale residuals are inflated close to 1 by the logarithmic transformation, while the large negative values are shrunk by the square root (Therneau et al., 1990). Plot of deviance residuals can be used to assess the prediction accuracy of failure rate for a given model (Fleming and Harrington, 1991).

2.4.2.3 Score residuals

Let $\hat{\beta} = (\hat{\beta}_1, \hat{\beta}_1, \ldots, \hat{\beta}_p)$ be a p-dimensional vector which maximizes the log-partial likelihood in (2.60), i.e. the root of the score function given in (2.61). Then the score residual of the ith subject and the jth covariate is defined as

$$r_{ij}(\hat{\beta},t) = \int_{x\in[0,t]} \left\{ Z_{ij}(x) - \frac{\sum_{i=1}^n H_i(x)\exp\left(\hat{\beta}'Z_i(x)\right)Z_{ij}(x)}{\sum_{i=1}^n H_i(x)\exp\left(\hat{\beta}'Z_i(x)\right)} \right\} d\widehat{M}_i(x), \tag{2.81}$$

where $i = 1, \ldots, n$, $j = 1, \ldots, p$, and $\widehat{M}_i(s)$ is the martingale residual defined in (2.73). The score residuals sum to zero according to the definition of $\hat{\beta}$.

The score residual r_{ij} given above can be interpreted as a weighted difference between value of the given covariate for the given subject and average value of this covariate in the risk set. It can be used in diagnosis of leverage of each subject on parameter estimates and in assessing proportional hazards assumptions.

2.4.3 *Discussion*

The Cox proportional hazard models are considered under left-truncation and right-censoring in this section. Comparing to models only under censoring, the difference is in the *risk set*. Under left-truncation and right-censoring, the risk set becomes $\tilde{X}_i \geq t > Y_i$, rather than $\tilde{X}_i \geq t$ (if only right-censoring is involved). If X is right-truncated by Y, the problem cannot be solved conveniently by the method in previous subsection, because we will not have (2.53) or (2.55) any more.

Consider the right-truncation case and we can only observe the subject information if $\tilde{X} = \min\{X, C\} \leq Y$. We have

$$\begin{aligned}
&\mathbf{P}(\tilde{X}_i \in [t,t+dt), \delta_i = 1 | \mathcal{G}_{i,\tau}, \tilde{X}_i \geq t, Y_i \geq t) \\
&= \mathbf{P}(\tilde{X}_i \in [t,t+dt), \delta_i = 1, Y_i \geq t | \mathcal{G}_{i,\tau}) / \mathbf{P}(\tilde{X}_i \geq t, Y_i \geq t | \mathcal{G}_{i,\tau}) \\
&= \mathbf{P}(X^*_{ji} \in [t,t+dt), C^*_{ji} \geq t, Y^*_{ji} \geq t | \mathcal{G}^*_{ji,\tau}, \tilde{X}^*_{ji} \leq Y^*_{ji}) / \mathbf{P}(\tilde{X}^*_{ji} \geq t, Y^*_{ji} \geq t | \mathcal{G}^*_{ji,\tau}, \tilde{X}^*_{ji} \leq Y^*_{ji}) \\
&= \mathbf{P}(X^*_{ji} \in [t,t+dt), C^*_{ji} \geq t, Y^*_{ji} \geq t | \mathcal{G}^*_{ji,\tau}) / \mathbf{P}(\tilde{X}^*_{ji} \geq t, \tilde{X}^*_{ji} \leq Y^*_{ji} | \mathcal{G}^*_{ji,\tau})
\end{aligned} \tag{2.82}$$

which cannot be simplified to the hazard rate function $d\Lambda_X(t|\mathcal{G}_{i,\tau})$. We need a little more effort here. If we assume that no censoring involved (or $C \geq Y$ almost surely). The above equation can be written as

$$\begin{aligned}
&\mathbf{P}(\tilde{X}_i \in [t,t+dt), \delta_i = 1|\mathcal{G}_{i,\tau}, \tilde{X}_i \geq t, Y_i \geq t) \\
&= \mathbf{P}(X^*_{j_i} \in [t,t+dt), Y^*_{j_i} \geq t|\mathcal{G}^*_{j_i,\tau})/\mathbf{P}(X^*_{j_i} \geq t, X^*_{j_i} \leq Y^*_{j_i}|\mathcal{G}^*_{j_i,\tau}) \\
&= d\Lambda_X(t|\mathcal{G}_{i,\tau})\frac{\mathbf{P}(Y \geq t|\mathcal{G}_{i,\tau})\mathbf{P}(X \geq t|\mathcal{G}_{i,\tau})}{\mathbf{P}(X \geq t, X \leq Y|\mathcal{G}_{i,\tau})} \\
&= d\Lambda_X(t|\mathcal{G}_{i,\tau})\frac{\mathbf{P}(Y \geq t)\mathbf{P}(X \geq t|\mathcal{G}_{i,\tau})}{\int_t^\tau [1 - F_Y(x|\mathcal{G}_{i,\tau})]dF_X(x|\mathcal{G}_{i,\tau})}. \qquad (2.83)
\end{aligned}$$

We may need an extra model to simplify the fraction term in (2.83). Or it may be easier to consider the retro-hazard (equivalent to consider the time reversal); see Example III.4.5 in Andersen et al. (1993).

The methods discussed in this chapter may also be extended to the mixture cure models (Yu and Peng, 2008), which are suitable for survival analysis with a cure fraction. In many medical and epidemiological studies, some patients may never experience the event of interest, because the patient is cured or immune to the disease. This could lead to very high percentage of censoring if we do not consider a cure model. For each subject, we will observe an indicator I, with $I = 1$ if an individual is cured and $I = 0$ otherwise. The probability of cure is modelled by a logistic regression

$$\mathbf{P}(I = 0|\boldsymbol{Z}) = \frac{1}{1 + \exp(\beta'\boldsymbol{Z})}$$

and the survival probability for the uncured patients modelled separately via $S(t|\boldsymbol{Z})$. Then the overall survival probability is given by $S_o(t|\boldsymbol{Z}) = \mathbf{P}(I = 0|\boldsymbol{Z}) + \mathbf{P}(I = 1|\boldsymbol{Z})S(t|\boldsymbol{Z})$. Even if $S(\infty|\boldsymbol{Z}) = 0$, the above mixture model implies that the main survival event T is such that $P(T = \infty|\boldsymbol{Z}) = \mathbf{P}(I = 0|\boldsymbol{Z})$. All the cured subjects will be censored subjects if we do not consider such cure models. Therefore for survival analysis with large censoring percentages at the last follow-up, such mixture cure models may still be useful.

Chapter 3
Bivariate estimation with truncated survival data

3.1 Introduction

Except for the univariate survival data discussed in the previous chapter, pairs of correlated event times, say (T_1, T_2), are also observed in many medical studies. For instance, in oncology studies when analysing time to cancer diagnosis in left and right breasts or in a paediatric AIDS study when the interests are incubation time of the mother and time from birth to AIDS development for the child. Estimation of their joint distribution or survival function is then of interest because it is useful in prediction of the joint survival experience and estimation of the correlation between the pairwise event times. There are many different types of dependence structures between two survival event times, including simple linear relationship, nonlinear relationship where the two event times may be modelled via a nonlinear regression model, early or late dependence where the dependence of the two events only occur when the event times are very small or very large, and causal relation where one event occurs shortly after the other event. Therefore, nonparametric (or at least semiparametric) modelling for bivariate survival analysis is usually preferable to parametric modelling in practice, since nonparametric methods make fewer assumptions and have a much wider applicability especially when very less information is known for the correlation structure.

Before introducing more existing research works of nonparametric bivariate survival function estimation under right censoring, we first discuss the copula model, which is very popular in high-dimensional statistical applications because a high dimensional cumulative distribution function F can be decomposed into two parts: the marginal distributions and the copula function C. Therefore, in statistical applications we can deal with the marginal survival functions first and then deal with the joint behavior of the pair of variables by taking into account the marginal distributions via a copulas. In a two-dimensional case, we can write

$$\begin{aligned} F(t_1, t_2) &= \mathbf{P}(F_1(T_1) \leq F_1(t_1), F_2(T_2) \leq F_2(t_2)) \\ &= F(F_1^{-1}(F_1(t_1)), F_2^{-1}(F_2(t_2))), \end{aligned}$$

where F_j^{-1} means

$$F_j^{-1}(t) = \inf\{x : F_j(x) \geq t\}.$$

If we denote the copula $C(u_1, u_2) = F(F_1^{-1}(u_1), F_2^{-1}(u_2))$, then the bivariate distribution function can be written as $F(t_1, t_2) = C(F_1(t_1), F_2(t_2))$. When using the copula method in practice, the copula function can be chosen from a collection of parametric functions. The estimation of the parameters in the copula can then be treated via parametric statistical methods. For example see Clayton (1978), Hougaard (1986) and Oakes (1989).

The selection of the parametric copula function may limit the application of copula models. In this book we focus on the more reliable nonparametric estimation of the joint survival functions. To estimate

Analysis for Time-to-Event Data under Censoring and Truncation.
http://dx.doi.org/10.1016/B978-0-12-805480-2.50003-4,

the bivariate distribution function $F(t_1,t_2) = \mathbf{P}(T_1 \leq t_1, T_2 \leq t_2)$, the challenge is mainly from the incompletely observed data. For example, (T_1, T_2) may be subject to random right censoring by a pair of censoring times (C_1, C_2). There is a vast literature for estimation of the joint survival function under right censoring, for example, Campbell (1981), Burke (1988), Dabrowska (1988), Dabrowska (1989), Dai and Bao (2009), Lin and Ying (1993), Prentice et al. (2004), Tsai et al. (1990), van der Laan (1996a) and Akritas and Keilegom (2003). When a single component of the paired event times is subject to truncation, Gürler (1996, 1997) proposed a nonparametric estimator. For nonparametric estimation of doubly truncated survival data, see for example van der Laan (1996b) and Huang et al. (2001). In the presence of both censoring and truncation, Gijbels and Gürler (1998) proposed a useful nonparametric estimator for the bivariate distribution function, however their method confines to the special case where a single component of the bivariate event times is subject to both censoring and truncation but the other one can be fully observed. Many of these estimators are not proper probability distributions or depend heavily on the choice of smoothing parameters or are only valid under very specific conditions.

The challenge of bivariate survival function estimation under censoring can be explained by that there is no unique way to reallocate the weights of censoring observations. Recall the univariate KM estimator. The univariate KM estimator only has jumps at uncensored values and assigns 0 weight on censored values. This is equivalent to allocating the weight of each censored value to all those uncensored values, which are larger than the censored value (those are still at risk), and then estimate the survival function based on the reweighted uncensored values. The same idea, however, cannot be easily applied to the bivariate case. This is because for bivariate survival analysis, we actually consider the estimate on a plane. If both events times are censored, there are many different ways to reallocate the weights in the plane. It is even more challenging to deal with subjects with one event time observed but the other one censored. In such cases, there is no obvious way to reallocate its weight. Each of the existing methods seems to provide an approach to reallocate the weight of censored observations, however none of them is uniformly better than the others.

If truncation is also involved, the nonparametric estimation of bivariate survival function will become more complicated, as we have seen in Chapter 2 that truncation is generally more difficult to be dealt with than censoring. In this chapter, we provide a general review of the nonparametric estimation for bivariate survival functions under both censoring and truncation.

3.2 Bivariate distributions

Let (T_1, T_2) be the pair of event times with survival function $S(t_1,t_2) = \mathbf{P}(T_1 > t_1, T_2 > t_2)$. Its relation with the bivariate distribution function $F(t_1,t_2) = \mathbf{P}(T_1 \leq t_1, T_2 \leq t_2)$ is given by $F(t_1,t_2) = 1 - S(t_1,0) - S(0,t_2) + S(t_1,t_2)$. The bivariate hazard function is defined as

$$\lambda(t_1,t_2) = \frac{d^2 S(t_1,t_2)/dt_1 dt_2}{S(t_1,t_2)}$$

or the more general formula (if S allows atom mass at certain points)

$$\Lambda(dt_1,dt_2) = \frac{\mathbf{P}(T_1 \in dt_1, T_2 \in dt_2)}{S(t_1-,t_2-)},$$

which means the probability that both coordinates will experience an event at (t_1,t_2) conditional on that they are at risk.

Note that a bivariate function S must satisfy the following conditions to be a survival function (i.e. can induce a probability measure):

Condition 3.2.1 *1. $S(u,v)$ is a decreasing function for $u \geq 0, v \geq 0$, i.e. if $(u_1,v_1) \leq (u_2,v_2)$ then $S(u_1,v_1) \geq S(u_2,v_2)$. Here $(u_1,v_1) \leq (u_2,v_2)$ is defined as $u_1 \leq u_2$ and $v_1 \leq v_2$.*
2. $S(u,v) \in [0,1]$ for $u \geq 0, v \geq 0$.

3. For any $u \geq 0, v \geq 0$*, we have* $S(u-,v-) - S(u-,v) - S(u,v-) + S(u,v) \geq 0$.

The relation of $S(t_1,t_2)$ and $\Lambda(t_1,t_2) = \int_{[0,t_1]}\int_{[0,t_2]} \lambda(u,v)dudv$ is more complicated than the univariate case relation. If S is continuous, we have

$$\begin{aligned}
\Lambda(t_1,t_2) &= \int_0^{t_1}\int_0^{t_2} \frac{d^2S(u,v)/dudv}{S(u,v)}dudv \\
&= \int_0^{t_1}\left[\frac{d}{du}\int_0^{t_2}\frac{d\log S(u,v)}{dv}dv\right]du + \int_0^{t_1}\int_0^{t_2}\frac{\frac{dS}{dv}\frac{dS}{du}}{S(u,v)^2}dudv \\
&= \int_0^{t_1}\left[\frac{d}{du}[\log S(u,t_2) - \log S(u,0)]\right]du + \int_0^{t_1}\int_0^{t_2}\frac{\frac{dS}{dv}\frac{dS}{du}}{S(u,v)^2}dudv \\
&= \log S(t_1,t_2) - \log S(0,t_2) - \log S(t_1,0) + \log S(0,0) + \int_0^{t_1}\int_0^{t_2}\frac{\frac{dS}{dv}\frac{dS}{du}}{S(u,v)^2}dudv,
\end{aligned}$$

where the last term $\int_0^{t_1}\int_0^{t_2}\frac{\frac{dS}{dv}\frac{dS}{du}}{S(u,v)^2}dudv$ cannot be represented via a simple term of either Λ or S. The above equation, however, implies that

$$\log S(t_1,t_2) - \log S(0,t_2) - \log S(t_1,0) + \log S(0,0) = \int_0^{t_1}\int_0^{t_2}\frac{d^2\log S(u,v)}{dudv}dudv.$$

In general (even for S not being continuous), with $S(0,0) = 1$, we can write

$$S(t_1,t_2) = S(0,t_2)\cdot S(t_1,0)\cdot \exp\left(\int_{[0,t_1]}\int_{[0,t_2]}(\log S)(du,dv)\right). \tag{3.1}$$

The above integration $\int_{[0,t_1]}\int_{[0,t_2]}(\log S)(du,dv)$ is a two-dimensional Lebesgue-Stieltjes integration, which is much more complicated than one-dimensional case. Within the integration range $[0,t_1]\times[0,t_2]$ we need to consider four types of points (u,v) with different continuity property. Define

$$\begin{aligned}
\mathscr{E}_1 &= \{(u,v) : (\log S)(\{u\},v) = (\log S)(u,\{v\}) = 0\}, \\
\mathscr{E}_2 &= \{(u,v) : (\log S)(\{u\},v) < 0, (\log S)(\{u\},\{v\}) = 0\}, \\
\mathscr{E}_3 &= \{(u,v) : (\log S)(\{u\},v) = 0, (\log S)(\{u\},\{v\}) < 0\}, \\
\mathscr{E}_4 &= \{(u,v) : (\log S)(\{u\},\{v\}) > 0\},
\end{aligned} \tag{3.2}$$

where for a function h, $h(\{x\}) = h(x) - h(x-)$. In the above definition, $\mathscr{E}_1$ is the set of continuous points and $\mathscr{E}_4$ is the set of discrete points. The sets $\mathscr{E}_2$ and $\mathscr{E}_3$ are points which are continuous in one coordinate but discrete in the other.

We can then write

$$\int_{[0,t_1]}\int_{[0,t_2]}(\log S)(du,dv) = \sum_{i=1}^{4} A_i(t_1,t_2),$$

where

$$\begin{aligned}
A_1(t_1,t_2) &= \int_{[0,t_1]}\int_{[0,t_2]} I[(u,v)\in\mathscr{E}_1](\log S)(du,dv), \\
A_2(t_1,t_2) &= \sum_{u\leq t_1}\int_{[0,t_2]} I[(u,v)\in\mathscr{E}_2](\log S)(\{u\},dv), \\
A_3(t_1,t_2) &= \sum_{v\leq t_2}\int_{[0,t_2]} I[(u,v)\in\mathscr{E}_3](\log S)(du,\{v\}), \\
A_4(t_1,t_2) &= \sum_{u\leq t_1}\sum_{v\leq t_2} I[(u,v)\in\mathscr{E}_4](\log S)(\{u\},\{v\}).
\end{aligned}$$

In bivariate survival analysis, one may be interested in not only the cumulative hazard rate $\Lambda(t_1,t_2)$, but also the following *conditional* hazard functions,

$$\Lambda_{10}(dt_1,t_2) = \frac{\mathbf{P}(T_1 \in dt_1, T_2 > t_2)}{\mathbf{P}(T_1 \geq t_1, T_2 > t_2)},$$
$$\Lambda_{01}(dt_1,t_2) = \frac{\mathbf{P}(T_1 > t_1, T_2 \in dt_2)}{\mathbf{P}(T_1 > t_1, T_2 \geq t_2)}. \tag{3.3}$$

The following relation between $(\Lambda,\Lambda_{10},\Lambda_{01})$ and S is very useful in practice,

$$\frac{S(t_1,t_2)}{S(t_1-,t_2-)} = 1-\Lambda_{10}(\{t_1\},t_2-)-\Lambda_{10}(t_1-,\{t_2\})+\Lambda_{11}(\{t_1\},\{t_2\}),$$
$$\frac{S(t_1-,t_2)}{S(t_1-,t_2-)} = 1-\Lambda_{01}(t_1-,\{t_2\}),$$
$$\frac{S(t_1,t_2-)}{S(t_1-,t_2-)} = 1-\Lambda_{10}(\{t_1\},t_2-). \tag{3.4}$$

If we define function $L(t_1,t_2)$ as

$$L(du,dv) = \frac{\Lambda_{10}(du,v-)\Lambda_{01}(u-,dv)-\Lambda(du,dv)}{[1-\Lambda_{10}(\{u\},v-)][1-\Lambda_{01}(u-,\{v\})]}, \tag{3.5}$$

it can be shown that

$$S(t_1,t_2) = S(0,t_2)\cdot S(t_1,0)\cdot \prod_{\substack{u\in[0,t_1]\\ v\in[0,t_2]}} [1-L(du,dv)]. \tag{3.6}$$

Based on the above relations, Dabrowska (1988) proposed a bivariate survival function estimator for censored observations. The above relations also imply the difficulties of finding a nonparametric estimate for the bivariate survival functions.

3.3 Types of bivariate truncated survival data

Let (T_1,T_2) be the pair of event times, which may be subject to random right censoring by a pair of censoring times (C_1,C_2). Thus we can only observe $Y_k = \min(T_k,C_k)$ and $\delta_k = I[T_k \leq C_k]$, $k=1,2$. The observation $(Y_1,Y_2,\delta_1,\delta_2)$ is subject to left truncation by a pair of truncation variables, denoted as (L_1,L_2). We assume that (T_1,T_2) to be independent of (C_1,C_2,L_1,L_2). In practice, there could be two different types of truncation.

3.3.1 *Type-A left truncation*

In some practical problems, truncation occurs when only subjects with $L_1 < Y_1$ *and* $L_2 < Y_2$ are observed. In other words we can observe $(Y_1,Y_2,\delta_1,\delta_2,L_1,L_2)$ only if $L_k < Y_k$, $k=1,2$ and nothing otherwise.

Example 3.1. (Two truncation times and no censoring time.) Huang et al. (2001) studied a bioplar affective disorder data where the bivariate event times of interest were ages onset of several parent-child pairs. The pairwise ages on set were both right truncated at their respective ages at interview since for an affected parent-child pair to be included in the study, they had to be diagnosed with the disease before their interviews. This is an example of doubly right truncated survival data in the absence of censoring, i.e. $C_1 = \infty$ and $C_2 = \infty$.

Example 3.2. (Two truncation times and one censoring time.) In the paediatric AIDS data described in Example 1.4, the incubation time of the mother and the time from birth to development of AIDS for the child constituted a bivariate survival data. The bivariate event times were left truncated if onset of AIDS occurred before the start of recruitment. After recruitment, the durations from birth to development of AIDS for all the children were fully observed but some of the incubation times for the mothers were right censored. This is a special case of type-A truncation where both event times are subject to random left truncation and only one is subject to random right censoring, i.e. $C_2 = \infty$.

3.3.2 Type-B left truncation

Sometimes, truncation occurs when only subjects with $L_1 < Y_1$ *or* $L_2 < Y_2$ are observed. In other words we can observe $(Y_1, Y_2, \delta_1, \delta_2, L_1, L_2)$ if $L_1 < Y_1$ *or* $L_2 < Y_2$ and nothing otherwise.

Example 3.3. In oncology studies when analysing time to cancer detection in the breast, a patient will be included in the study if cancer is detected in either left or right breast within the study period, i.e. truncation occurs if cancer has already been detected in either left or right breast before the start of the study.

3.3.3 Other types of truncation

As mentioned in Chapter 2, there are different types of univariate truncations, such as right truncation, interval truncation and length-bias. All these different types make the bivariate survival analysis very complicated, since the two survival times T_1 and T_2 may be subject to different types of truncations. In this chapter, we mainly focus on left truncation and right censoring.

3.4 The inverse probability weighted estimator with only one censoring variable

For both survival times subject to random left truncation and one of the survival times subject to random right censoring, Shen (2006) proposed an inverse-probability-weighted (IPW) approach to estimate the bivariate distribution and survival function (see Example 3.2). In this section, we will mainly discuss the estimation method proposed by Shen (2006). The estimators of van der Laan (1996b) and Gijbels and Gürler (1998) are introduced as special cases of Shen's IPW estimator.

The bivariate distribution function and survival function of (T_1, T_2) are denoted by, respectively,

$$\begin{aligned} F(t_1,t_2) &= \mathbf{P}(T_1 \le t_1, T_2 \le t_2), \\ S(t_1,t_2) &= \mathbf{P}(T_1 > t_1, T_2 > t_2) = 1 - F(t_1,\infty) - F(\infty,t_2) + F(t_1,t_2). \end{aligned} \tag{3.7}$$

We consider the scenario of Example 3.2, where there is only one censoring variable C, corresponding to T_1, i.e. $\delta = I[T_1 \le C]$ and $Y = \min(T_1, C)$. The other event time T_2 is not censored (corresponding to $C_1 = C$ and $C_2 = \infty$ in Example 3.2). The (type-A) truncation variables are denoted as (L_1, L_2). We assume that (T_1, T_2) is independent of (L_1, L_2, C), but the truncation variables (L_1, L_2) and the censoring variable C may be correlated.

To develop the IPW estimator for F, we also need to define $G_L(t_1,t_2) = \mathbf{P}(L_1 \le t_1, L_2 \le t_2)$ and $Q(t_1,t_2) = \mathbf{P}(C \le t_1, L_2 \le t_2)$ as the joint distribution function of (L_1, L_2) and (C, L_2), respectively. Note that for this special example we do not need the joint distribution for the triple (L_1, L_2, C). Similarly as before, we use letters with subscript i to denote the observed truncated data, i.e. the observed data are denoted by $(L_{1i}, L_{2i}, Y_i, \delta_i, T_{2i})$, $i = 1, \ldots, n$.

Consider the sub-distribution function

$$\begin{aligned} F^*(t_1,t_2) &= \mathbf{P}(Y_i \le t_1, \delta_i = 1, T_{2i} \le t_2) \\ &= \mathbf{P}(T_1 \le t_1, T_1 \le C, T_2 \le t_2 | L_1 < T_1, L_2 < T_2) \\ &= \gamma^{-1}\mathbf{P}(T_1 \le t_1, L_1 < T_1 \le C, L_2 < T_2 \le t_2) \\ &= \gamma^{-1} \int_{[0,t_2]} \int_{[0,t_1]} [G_L(u-,v-) - Q(u-,v-)] F(du,dv), \end{aligned} \tag{3.8}$$

where $\gamma = \mathbf{P}(L_1 < T_1, L_2 < T_2)$ denotes the truncation probability. Thus, we have

$$F(dt_1,dt_2) = \gamma \cdot \frac{F^*(dt_1,dt_2)}{G_L(t_1-,t_2-) - Q(t_1-,t_2-)}.$$

Since $F^*(dt_1,dt_2)$ can be estimated via the empirical distribution function $\hat{F}^*(dt_1,dt_2) = n^{-1}\sum_{i=1}^n I[Y_i \in [t_1,t_1+dt_1), \delta_i = 1, T_{2i} \in [t_2,t_2+dt_2)]$, we have that when G_L, Q and γ are all known $F(t_1,t_2)$ can be estimated by

$$\begin{aligned} &\gamma \cdot \int_{[0,t_2]} \int_{[0,t_1]} n^{-1} \sum_{i=1}^n \frac{I[Y_i \in [u,u+du), T_{2i} \in [v,v+dv)]\delta_i}{G_L(u-,v-) - Q(u-,v-)} \\ &= n^{-1}\gamma \cdot \sum_{i=1}^n \frac{I[Y_i \le t_1, T_{2i} \le t_2]\delta_i}{G_L(Y_i-,T_{2i}-) - Q(Y_i-,T_{2i}-)}. \end{aligned} \tag{3.9}$$

Since the above cumulative distribution estimator should be equal to 1 at point $t_1 = t_2 = \infty$, the truncation probability γ can be estimated by

$$\hat{\gamma} = n \cdot \left[\sum_{i=1}^n \frac{\delta_i}{G_L(Y_i-,T_{2i}-) - Q(Y_i-,T_{2i}-)} \right]^{-1}.$$

Thus, when G_L and Q are known, $F(t_1,t_2)$ can be estimated by

$$\left[\sum_{i=1}^n \frac{\delta_i}{G_L(Y_i-,T_{2i}-) - Q(Y_i-,T_{2i}-)} \right]^{-1} \sum_{i=1}^n \frac{I[Y_i \le t_1, T_{2i} \le t_2]\delta_i}{G_L(Y_i-,T_{2i}-) - Q(Y_i-,T_{2i}-)}.$$

Next, consider the sub-distribution function $G^*(t_1,t_2) = \mathbf{P}(L_{1i} \le t_1, L_{2i} \le t_2)$ and we have the following relationship

$$\begin{aligned} G^*(t_1,t_2) &= \mathbf{P}(L_1 \le t_1, L_2 \le t_2 | L_1 < T_1, L_2 < T_2) \\ &= \gamma^{-1} \int_{[0,t_2]} \int_{[0,t_1]} S(u,v) G_L(du,dv), \end{aligned} \tag{3.10}$$

Similarly, when $S(t_1,t_2)$ and γ are known, $G_L(t_1,t_2)$ can be estimated by

$$n^{-1}\gamma \cdot \sum_{i=1}^n \frac{I[L_{1i} \le t_1, L_{2i} \le t_2]}{S(L_{1i}, L_{2i})}.$$

Again, the above cumulative distribution estimator should be 1 at point $t_1 = t_2 = \infty$. Therefore the truncation probability γ can also be estimated by

$$\hat{\gamma} = n \cdot \left[\sum_{i=1}^{n} \frac{1}{S(L_{1i}, L_{2i})} \right]^{-1}.$$

Hence, when $S(t_1, t_2)$ is known, $G_L(t_1, t_2)$ can be estimated by

$$\left[\sum_{i=1}^{n} \frac{1}{S(L_{1i}, L_{2i})} \right]^{-1} \sum_{i=1}^{n} \frac{I[L_{1i} \le t_1, L_{2i} \le t_2]}{S(L_{1i}, L_{2i})}.$$

Similarly, consider the sub-distribution function $Q^*(t_1, t_2) = \mathbf{P}(Y_i \le t_1, \delta_i = 0, L_{2i} \le t_2)$ and we have the following relationship, under the assumption $\mathbf{P}(L_1 \le C) = 1$,

$$\begin{aligned} Q^*(t_1, t_2) &= \mathbf{P}(C \le t_1, C < T_1, L_2 \le t_2 | L_1 < T_1, L_2 < T_2) \\ &= \gamma^{-1} \int_{[0,t_2]} \int_{[0,t_1]} S(u, v) Q(du, dv), \end{aligned} \tag{3.11}$$

When $S(t_1, t_2)$ is known, $Q(t_1, t_2)$ can be estimated by

$$\left[\sum_{i=1}^{n} \frac{1-\delta_i}{S(Y_i, L_{2i})} \right]^{-1} \sum_{i=1}^{n} \frac{I[Y_i \le t_1, L_{2i} \le t_2](1-\delta_i)}{S(Y_i, L_{2i})}.$$

According to the arguments above, the IPW estimators of $F(t_1, t_2)$, $G_L(t_1, t_2)$ and $Q(t_1, t_2)$ can be obtained by simultaneously estimating the following three distribution functions:

$$\hat{F}(t_1, t_2) = \left[\sum_{i=1}^{n} \frac{\delta_i}{\hat{G}_L(Y_i-, T_{2i}-) - \hat{Q}(Y_i-, T_{2i}-)} \right]^{-1} \sum_{i=1}^{n} \frac{I[Y_i \le t_1, T_{2i} \le t_2]\delta_i}{\hat{G}_L(Y_i-, T_{2i}-) - \hat{Q}(Y_i-, T_{2i}-)}, \tag{3.12}$$

$$\hat{G}_L(t_1, t_2) = \left[\sum_{i=1}^{n} \frac{1}{\hat{S}(L_{1i}, L_{2i})} \right]^{-1} \sum_{i=1}^{n} \frac{I[L_{1i} \le t_1, L_{2i} \le t_2]}{\hat{S}(L_{1i}, L_{2i})}, \tag{3.13}$$

$$\hat{Q}(t_1, t_2) = \left[\sum_{i=1}^{n} \frac{1-\delta_i}{\hat{S}(Y_i, L_{2i})} \right]^{-1} \sum_{i=1}^{n} \frac{I[Y_i \le t_1, L_{2i} \le t_2](1-\delta_i)}{\hat{S}(Y_i, L_{2i})}. \tag{3.14}$$

In practice, one can choose an initial estimator for $F(t_1, t_2)$, denoted by $\hat{F}^{(0)}(t_1, t_2)$, then iterate the following steps until convergence is established.

1. Calculate the corresponding $\hat{S}^{(m)}(t_1, t_2)$, based on $\hat{F}^{(m)}(t_1, t_2)$.
2. Calculate $\hat{G}_L^{(m)}(t_1, t_2)$ and $\hat{Q}^{(m)}(t_1, t_2)$ based on $\hat{S}^{(m)}(t_1, t_2)$, (3.13) and (3.14).
3. Update the estimators $\hat{F}^{(m+1)}(t_1, t_2)$ based on (3.12).

Shen (2006) showed that his IPW estimators were consistent under certain conditions, based on Theorem 3.1 of van der Laan (1996b). He also established simulation studies which indicated the IPW estimators worked satisfactorily for moderate sample size. However, analytic expressions of the asymptotic variances of the estimators were not given in his paper.

It can be proved that under some circumstances, Shen's estimator reduces to the NPMLE proposed by van der Laan (1996b) or the estimator of Gijbels and Gürler (1998).

1. *Special case (i): there is no right censoring, i.e.* $C=\infty$.

When $C=\infty$, it follows that $Q(t_1,t_2)=0$ and (3.12) reduces to

$$\hat{F}(t_1,t_2)=\left[\sum_{i=1}^{n}\frac{1}{\hat{G}_L(T_{1i}-,T_{2i}-)}\right]^{-1}\sum_{i=1}^{n}\frac{I[T_{1i}\le t_1,T_{2i}\le t_2]}{\hat{G}_L(T_{1i}-,T_{2i}-)}. \tag{3.15}$$

In this case, Shen (2006) has proved the equivalence of (3.15) and the NPMLE of $F(t_1,t_2)$ proposed by van der Laan (1996b) which was of the form

$$\hat{F}_n(t_1,t_2)=\sum_{i=1}^{n}\frac{I[T_{1i}\le t_1,T_{2i}\le t_2]}{\sum_{j=1}^{n}\frac{I[L_{1j}\le T_{1i},L_{2j}\le T_{2i}]}{\hat{S}_n(L_{1j}-,L_{2j}-)}}. \tag{3.16}$$

2. *Special case (ii): there is random left truncation and right censoring on only one of the event times, i.e.* $L_2=0$.

When $L_2=0$, the bivariate distribution functions G_L and Q reduce to the univariate forms $\tilde{G}(t)=\mathbf{P}(L\le t)$ and $\tilde{Q}(t)=\mathbf{P}(C\le t)$. The observed data now are denoted by $(L_i,Y_i,\delta_i,T_{2i})$, $i=1,\ldots,n$. Following the idea of Shen's IPW estimator, the sub-distribution function

$$\begin{aligned}F^*(t_1,t_2)&=\mathbf{P}(Y_i\le t_1,\delta_i=1,T_{2i}\le t_2)\\&=\mathbf{P}(T_1\le t_1,T_1\le C,T_2\le t_2|L<T_1)\\&=\tilde{\gamma}^{-1}\mathbf{P}(T_1\le t_1,T_2\le t_2,L<T_1\le C)\\&=\tilde{\gamma}^{-1}\int_{[0,t_2]}\int_{[0,t_1]}[\tilde{G}(u-)-\tilde{Q}(u-)]F(du,dv),\end{aligned}$$

where $\tilde{\gamma}=\mathbf{P}(L\le T_1)$ denotes the truncation probability, or equivalently

$$F(t_1,t_2)=\tilde{\gamma}\int_{[0,t_2]}\int_{[0,t_1]}[\tilde{G}(u-)-\tilde{Q}(u-)]^{-1}F^*(du,dv),$$

Therefore when $\tilde{G}$, $\tilde{Q}$ and $\tilde{\gamma}$ are known, (3.9) reduces to

$$n^{-1}\tilde{\gamma}\cdot\sum_{i=1}^{n}\frac{I[Y_i\le t_1,T_{2i}\le t_2]\delta_i}{\tilde{G}(Y_i-)-\tilde{Q}(Y_i-)} \tag{3.17}$$

which is an estimate of $F(t_1,t_2)$.
Let

$$\begin{aligned}R(t)&=\mathbf{P}(L_i<t\le Y_i)\\&=\mathbf{P}(L<t\le C,t\le T_1|L\le T_1)\\&=\tilde{\gamma}^{-1}[\tilde{G}(t-)-\tilde{Q}(t-)]S(t-,0).\end{aligned} \tag{3.18}$$

Hence the expressions in (3.17) and (3.18) lead to the estimator: with $\hat{S}(t,0)$ as the univariate product-limit estimator for the marginal survival function $S(t,0)$ of T_1,

$$\begin{aligned}\hat{F}_n(t_1,t_2)&=n^{-1}\sum_{i=1}^{n}\frac{I[Y_i<t_1,T_{2i}\le t_2]\cdot\delta_i\cdot\hat{S}(Y_i-,0)}{\hat{R}(Y_i)},\\ \hat{R}(t)&=n^{-1}\sum_{i}I[L_i<t\le Y_i]\end{aligned} \tag{3.19}$$

which has the same form as the estimator proposed by Gijbels and Gürler (1998).

The estimator discussed in this section used the condition $\mathbf{P}(L_1 < C) = 1$, which corresponding to Condition 2.1.4 in Chapter 2. This is because without this condition, $\mathbf{P}(L_1 < u \leq C, L_2 < v) \neq G_L(u-, v-) - Q(u-, v-)$ and equation (3.8) will not hold and the above iteration algorithm fails. This condition, however, may not be reasonable in practice. The method introduced in the following section is not limited to such a condition and does not need an iteration algorithm.

3.5 The transformation estimator

In this section, we consider a more general case, where both survival times are subject to random right censoring and left truncation (type-A truncation). For such cases, Dai and Fu (2012) proposed an non-parametric estimator for the bivariate survival function. Their method was based on a polar coordinate transformation, which is briefly introduced in this section.

3.5.1 Notations and assumptions

Let (T_1, T_2) be the pair of event times. The bivariate distribution function and survival function of (T_1, T_2) are denoted by $F(t_1, t_2) = \mathbf{P}(T_1 \leq t_1, T_2 \leq t_2)$ and $S(t_1, t_2) = \mathbf{P}(T_1 > t_1, T_2 > t_2)$, respectively. The paired survival times (T_1, T_2) may be left truncated by a pair of truncation times (L_1, L_2) and right censored by a pair of censoring times (C_1, C_2). Thus one can only observe $Y_k = \min(T_k, C_k)$ and $\delta_k = I[T_k \leq C_k]$ when $L_k \leq T_k$, $k = 1, 2$. Dai and Fu (2012) also made the following assumptions throughout their paper.

Assumption 3.5.1 *(T_1, T_2) and (L_1, L_2, C_1, C_2) are independent, and the distribution $F(t_1, t_2)$ has a continuous support.*

Note that the censoring variables C and the truncation variables L are correlated, which is the main challenge in bivariate survival analysis under both censoring and truncation.

Assumption 3.5.2 *The lower boundaries of support for F are coordinate axes of the first quadrant.*

Assumption 3.5.3 *For $\mathbf{t} = (t_1, t_2)$, the function*

$$G(t_1, t_2) = \mathbf{P}(C_1 > t_1 \geq L_1, C_2 > t_2 \geq L_2) > 0. \tag{3.20}$$

almost surely with respect to $F(dt_1, dt_2)$ in $\mathscr{A}$, where $\mathscr{A}$ is the support area of F.

Note that Condition 3.5.3 corresponds to Condition 2.1.8 in Chapter 2 and it guarantees that $F(t_1, t_2)$ can be identified in the area under consideration (He and Yang, 1998).

3.5.2 Polar coordinate data transformation

Define a transformation from (t_1, t_2) to $(z; \alpha)$ as

$$\begin{aligned} z &= \sqrt{t_1^2 + t_2^2} \\ \alpha &= t_2 / t_1. \end{aligned} \tag{3.21}$$

Such a transformation is related to the polar coordinate transformation. Given $\theta \in [0, \pi/2]$, such that $\cos\theta = 1/\sqrt{1+\alpha^2}$ or $\sin\theta = 1/\sqrt{1+\alpha^{-2}}$, one have that $t_1 = z\cos\theta$ and $t_2 = z\sin\theta$. Hence (z, θ) is the polar coordinate of (t_1,t_2). Similarly we may also use notation (v_1,v_2) and its transformation $s=\sqrt{v_1^2+v_2^2}$ and $\beta=v_2/v_1$.

For fixed α, $S(t_1,t_2)$ can be transformed to a function, $S(z;\alpha)$, by the following formula,

$$S(t_1,t_2)=\mathbf{P}(T_1>t_1,T_2>t_2)=\mathbf{P}(Z(\alpha)>z):=S(z;\alpha), \tag{3.22}$$

where

$$Z(\alpha)=\min\left\{T_1\sqrt{1+\alpha^2},T_2\sqrt{1+\alpha^{-2}}\right\}. \tag{3.23}$$

It can be viewed as a univariate function if α is given.

Similarly, we can write

$$\begin{aligned} G(z;\alpha) &= G(t_1,t_2)\\ &= \mathbf{P}\left(\min\left\{C_1\sqrt{1+\alpha^2},C_2\sqrt{1+\alpha^{-2}}\right\}>z\geq\min\left\{L_1\sqrt{1+\alpha^2},L_2\sqrt{1+\alpha^{-2}}\right\}\right). \end{aligned} \tag{3.24}$$

In practice, the values of $Z(\alpha)$ in (3.23) may not be obtained due to censoring and truncation. In the presence of only censoring, with the same notation $Y_k=\min\{T_k,C_k\}$ as the previous section, define $\tilde{Y}_1=Y_1\sqrt{1+\alpha^2}$ and $\tilde{Y}_2=Y_2\sqrt{1+\alpha^{-2}}$. Then the transformed data without truncation are of the form

$$\begin{aligned} \tilde{Z}(\alpha) &= \min\{\tilde{Y}_1,\tilde{Y}_2\},\\ \Delta(\alpha) &= \delta_1 I[\tilde{Y}_1<\tilde{Y}_2]+\delta_2 I[\tilde{Y}_1>\tilde{Y}_2]+\max(\delta_1,\delta_2)I[\tilde{Y}_1=\tilde{Y}_2], \end{aligned} \tag{3.25}$$

We can show the following.

Lemma 3.1. *Based on the $Z(\alpha)$ given in (3.23) and the transformed data given in (3.25), if $\Delta(\alpha)=1$ then $\tilde{Z}(\alpha)=Z(\alpha)$ and if $\Delta(\alpha)=0$ then $\tilde{Z}(\alpha)<Z(\alpha)$.*

Proof. The lemma follows from the results shown below.

1. If $\delta_1=1$ and $\delta_2=1$, then $\Delta(\alpha)=1$, $\tilde{Y}_1=T_1\sqrt{1+\alpha^2}$ and $\tilde{Y}_2=T_2\sqrt{1+\alpha^{-2}}$ which indicate $\tilde{Z}(\alpha)=Z(\alpha)$.
2. If $\delta_1=0$ and $\delta_2=0$, then $\Delta(\alpha)=0$, $\tilde{Y}_1=C_1\sqrt{1+\alpha^2}<T_1\sqrt{1+\alpha^2}$ and $\tilde{Y}_2=C_2\sqrt{1+\alpha^{-2}}<T_2\sqrt{1+\alpha^{-2}}$ which indicate $\tilde{Z}(\alpha)<Z(\alpha)$.
3. If $\delta_1=1$ and $\delta_2=0$, then $\Delta(\alpha)=I[\tilde{Y}_1<\tilde{Y}_2]+I[\tilde{Y}_1=\tilde{Y}_2]$. Hence,
   - if $\tilde{Y}_1<\tilde{Y}_2$, then $\Delta(\alpha)=1$ and $\tilde{Z}(\alpha)=Z(\alpha)$;
   - if $\tilde{Y}_1=\tilde{Y}_2$, then $\Delta(\alpha)=1$ and $\tilde{Z}(\alpha)=Z(\alpha)$;
   - if $\tilde{Y}_1>\tilde{Y}_2$, then $\Delta(\alpha)=0$ and $\tilde{Z}(\alpha)<Z(\alpha)$.
4. If $\delta_1=0$ and $\delta_2=1$, similar results as those of case 3 can be obtained.

This implies that $\tilde{Z}(\alpha)$ is the censored observation for $Z(\alpha)$, and $\Delta(\alpha)$ is the indicator for censoring. Thus without truncation, for a given α the transformed univariate function $S(z;\alpha)$ can be estimated using the univariate Kaplan-Meier estimator based on the transformed data $\{\tilde{Z}(\alpha),\Delta(\alpha)\}$. Details can be found in Dai and Bao (2009).

In the presence of both censoring and truncation, one can observe $\{\tilde{Z}(\alpha),\Delta(\alpha)\}$ only if $L_1\leq T_1$ and $L_2\leq T_2$. Let $\tilde{Y}_{1,i}=Y_{1,i}\sqrt{1+\alpha^2}$ and $\tilde{Y}_{2,i}=Y_{2,i}\sqrt{1+\alpha^{-2}}$. The observed data after transformation are

$$\begin{aligned} \tilde{Z}_i(\alpha) &= \min\{\tilde{Y}_{1,i},\tilde{Y}_{2,i}\},\\ \Delta_i(\alpha) &= \delta_{1,i}I[\tilde{Y}_{1,i}<\tilde{Y}_{2,i}]+\delta_{2,i}I[\tilde{Y}_{1,i}>\tilde{Y}_{2,i}]+\max(\delta_{1,i},\delta_{2,i})I[\tilde{Y}_{1,i}=\tilde{Y}_{2,i}],\\ V_i(\alpha) &= \max\left\{L_{1,i}\sqrt{1+\alpha^2},L_{2,i}\sqrt{1+\alpha^{-2}}\right\}. \end{aligned} \tag{3.26}$$

Note that the truncation condition $\{L_1\leq Y_1,L_2\leq Y_2\}$ corresponds to the weaker Condition 2.1.5 for the univariate case. This condition is not equivalent to $\tilde{Z}(\alpha)\geq V(\alpha)$. Therefore the observed data after transformation $\{\tilde{Z}_i(\alpha),\Delta_i(\alpha),V_i(\alpha)\}$ is not exactly the same as the univariate left-truncated and right-censored

(LTRC) observation, as that in Chapter 2. We here need a little more work to extend the methodology in Chapter 2 to such transformed data.

3.5.3 Bivariate survival estimator

Based on the transformed observations $\{\tilde{Z}_i(\alpha), \Delta_i(\alpha), V_i(\alpha)\}$, $i = 1, \cdots, n$, (3.26), Dai and Fu (2012) defined

$$\begin{aligned} N(ds;\alpha) &= n^{-1}\sum_{i=1}^{n} N_i(ds;\alpha), \\ &:= n^{-1}\sum_{i=1}^{n} I\left[\tilde{Z}_i(\alpha) \in ds, s > V_i(\alpha), \Delta_i(\alpha) = 1\right], \quad (3.27) \\ H_{(n)}(s;\alpha) &= n^{-1}\sum_{i=1}^{n} H_i(s;\alpha), \\ &:= n^{-1}\sum_{i=1}^{n} I\left[\tilde{Z}_i(\alpha) > s \geq V_i(\alpha)\right], \quad (3.28) \\ H_{(n)}(t_1,t_2) &= n^{-1}\sum_{i=1}^{n} H_i(t_1,t_2), \\ &:= n^{-1}\sum_{i=1}^{n} I\left[Y_{1i} > t_1 \geq L_{1i}, Y_{2i} > t_2 \geq L_{2i}\right], \quad (3.29) \end{aligned}$$

where $Z \in ds$ denotes $s \leq Z < s + ds$ for simplicity and proved the following lemma which leads to the estimation of $S(z;\alpha)$.

Lemma 3.2. *For fixed α, the hazard rate function of $Z(\alpha)$ is denoted by $\Lambda(dz;\alpha) = -S(dz;\alpha)/S(z-;\alpha)$. Then we have*

$$\Lambda(dz;\alpha) = \frac{\mathbf{P}\left(\tilde{Z}_i(\alpha) \in dz, z > V_i(\alpha), \Delta_i(\alpha) = 1\right)}{\mathbf{P}\left(\tilde{Z}_i(\alpha) \geq z > V_i(\alpha)\right)}. \quad (3.30)$$

Proof. The definition of $\Lambda(dz;\alpha)$ means that

$$\Lambda(dz;\alpha) = \frac{\mathbf{P}(T_1 \in dt_1, T_2 > t_1) + \mathbf{P}(T_1 > t_1, T_2 \in dt_2) + \mathbf{P}(T_1 \in dt_1, T_2 \in dt_2)}{\mathbf{P}(T_1 \geq t_1, T_2 \geq t_2)}.$$

On the other hand

$$\begin{aligned} &\frac{\mathbf{P}\left(\tilde{Z}_i(\alpha) \in dz, z > V_i(\alpha), \Delta_i(\alpha) = 1\right)}{\mathbf{P}\left(\tilde{Z}_i(\alpha) \geq z > V_i(\alpha)\right)} \\ &= \frac{\mathbf{P}\left(\tilde{Z}(\alpha) \in dz, z > V(\alpha), \Delta(\alpha) = 1 | Y_1 < L_1, Y_2 < L_2\right)}{\mathbf{P}\left(\tilde{Z}(\alpha) \geq z > V(\alpha) | Y_1 < L_1, Y_2 < L_2\right)} \\ &= \frac{\mathbf{P}\left(\tilde{Z}(\alpha) \in dz, z > V(\alpha), \Delta(\alpha) = 1\right)}{\mathbf{P}\left(\tilde{Z}(\alpha) \geq z > V(\alpha)\right)}, \end{aligned}$$

where the second '=' is because $\{\tilde{Z}(\alpha) \geq z > V(\alpha)\} \subset \{Y_1 < L_1, Y_2 < L_2\}$.

The lemma is proved by noticing

$$
\begin{aligned}
&\mathbf{P}\left(\tilde{Z}(\alpha)\in dz, z>V(\alpha), \Delta(\alpha)=1\right)\\
&=\mathbf{P}(Y_1\in dt_1, Y_2>t_2, \delta_1=1, V(\alpha)<z)+\mathbf{P}(Y_1>t_1, Y_2\in dt_2, \delta_2=1, V(\alpha)<z)\\
&\quad+\mathbf{P}(Y_1\in dt_1, Y_2\in dt_2, \delta_1=1, \delta_2=0, V(\alpha)<z)+\mathbf{P}(Y_1\in dt_1, Y_2\in dt_2, \delta_1=0, \delta_2=1, V(\alpha)<z)\\
&\quad+\mathbf{P}(Y_1\in dt_1, Y_2\in dt_2, \delta_1=\delta_2=1, V(\alpha)<z)\\
&=\mathbf{P}(T_1\in dt_1, C_1\geq t_1, T_2>t_2, C_2>t_2, V(\alpha)<z)+\mathbf{P}(T_1>t_1, C_1>t_1, T_2\in dt_2, C_2\geq t_2, V(\alpha)<z)\\
&\quad+\mathbf{P}(T_1\in dt_1, C_1\geq t_1, C_2\in dt_2, T_2>t_2, V(\alpha)<z)+\mathbf{P}(C_1\in dt_1, T_2>t_1, T_2\in dt_2, C_2\geq t_2, V(\alpha)<z)\\
&\quad+\mathbf{P}(T_1\in dt_1, C_1\geq t_1, T_2\in dt_2, C_2\geq t_2, V(\alpha)<z)\\
&=\mathbf{P}(C_1\geq t_1, C_2\geq t_2, V(\alpha)<z)\{\mathbf{P}(T_1\in dt_1, T_2>t_2)+\mathbf{P}(T_1>t_1, T_2\in dt_2)+\mathbf{P}(T_1\in dt_1, T_2\in dt_2)\}
\end{aligned}
$$

and

$$\mathbf{P}\left(\tilde{Z}(\alpha)\geq z>V(\alpha)\right)=\mathbf{P}(C_1\geq t_1, C_2\geq t_2, V(\alpha)<z)\mathbf{P}(T_1\geq t_1, T_2\geq t_2).$$

■

Lemma 3.2 implies that an estimator for $\Lambda(dz;\alpha)$ is $\hat{\Lambda}(dz;\alpha)=N(dz;\alpha)/H_{(n)}(z-;\alpha)$. Then the product-limit estimator for $S(z;\alpha)$ proposed by Dai and Fu (2012) is

$$\hat{S}(z;\alpha)=\prod_{s\leq z}\left[1-\frac{N\{s;\alpha\}}{H_{(n)}(s-;\alpha)}\right], \tag{3.31}$$

where $N\{s;\alpha\}=N(s;\alpha)-N(s-;\alpha)=n^{-1}\sum_{i=1}^{n}I\left[\tilde{Z}_i(\alpha)=s, s>V_i(\alpha), \Delta_i(\alpha)=1\right]$. Since $S(z;\alpha)=S(t_1,t_2)$, $\hat{S}(z;\alpha)$ is also an estimator for $S(t_1,t_2)$.

Let

$$
\begin{aligned}
M_i(ds;\alpha)&=N_i(ds;\alpha)-H_i(s-;\alpha)\Lambda(ds;\alpha)\\
M(ds;\alpha)&=n^{-1}\sum_i M_i(ds;\alpha).
\end{aligned}
\tag{3.32}
$$

We also have

$$\hat{S}(z;\alpha)-S(z;\alpha)=-S(z;\alpha)\int_{[0,z]}\frac{\hat{S}(s-;\alpha)}{S(s;\alpha)}\frac{I[H_{(n)}(s-)>0]}{H_{(n)}(s-)}M(ds;\alpha)+B(z;\alpha) \tag{3.33}$$

with

$$B(z;\alpha)=S(z;\alpha)\int_{[0,z]}\frac{\hat{S}(s-;\alpha)}{S(s;\alpha)}I[H_{(n)}(s-)=0]\Lambda_X(ds;\alpha). \tag{3.34}$$

3.5.4 *Large sample properties for bivariate survival estimator*

For a given α, define σ-field

$$\mathscr{F}^i_{z;\alpha}=\sigma\left\{I[\tilde{Z}_i(\alpha)\leq s], \Delta_i(\alpha), I[V_i(\alpha)\leq s], 0\leq s\leq z\right\}. \tag{3.35}$$

Equipping the counting process $\{N_i(z;\alpha):z\geq 0\}$ with the filtration $\{\mathscr{F}^i_{z;\alpha}, z\geq 0\}$, we have

Lemma 3.3.

$$E\left[N_i(dz;\alpha)|\mathscr{F}^i_{z-;\alpha}\right]=H_i(z-;\alpha)\Lambda(dz;\alpha). \tag{3.36}$$

Proof. The proof is similar to that of Lemma 2.1. For simplicity of notations, we use $T \in dt$ to denote $T \in [t, t+dt)$.

We need to show that for any set $A \in \mathscr{F}^i_{z-;\alpha}$,

$$\int_A H_i(z-;\alpha)\Lambda(dz;\alpha)d\mathbf{P} = \int_A N_i(dz;\alpha)d\mathbf{P}. \tag{3.37}$$

(a) For set A of the form $\{\tilde{Z}_i(\alpha) \geq s_3, V_i(\alpha) \in [s_1, s_2)\}$ for any $0 \leq s_1 < s_2 < z$, $s_3 \leq z$, we have

$$\begin{aligned}
&\int_A H_i(z-;\alpha)\Lambda(dz;\alpha)d\mathbf{P} = \mathbf{P}(\tilde{Z}_i(\alpha) \geq z, V_i(\alpha) \in [s_1,s_2))\Lambda(dz;\alpha)\\
&= \mathbf{P}(Y_{1i} \geq t_1, Y_{2i} \geq t_2, V_i(\alpha) \in [s_1,s_2))\Lambda(dz;\alpha)\\
&= \mathbf{P}(Y_1 \geq t_1, Y_2 \geq t_2, V(\alpha) \in [s_1,s_2)|Y_1 > L_1, Y_2 > L_2)\Lambda(dz;\alpha)\\
&= \gamma^{-1}\mathbf{P}(T_1 \geq t_1, T_2 \geq t_2)\mathbf{P}(C_1 \geq t_1, C_2 \geq t_2, V(\alpha) \in [s_1,s_2))\Lambda(dz;\alpha)\\
&= \gamma^{-1}\mathbf{P}(C_1 \geq t_1, C_2 \geq t_2, V(\alpha) \in [s_1,s_2))\{\mathbf{P}(T_1 \in dt_1, T_2 > t_2) + \mathbf{P}(T_1 > t_1, T_2 \in dt_2) + \mathbf{P}(T_1 \in dt_1, T_2 \in dt_2)\}.
\end{aligned}$$

On the other hand

$$\begin{aligned}
&\int_A N_i(dz;\alpha)d\mathbf{P} = \mathbf{P}(\tilde{Z}_i(\alpha) \in dz, \Delta_i(\alpha) = 1, V_i(\alpha) \in [s_1,s_2))\\
&= \gamma^{-1}\mathbf{P}(\tilde{Z}(\alpha) \in dz, \Delta(\alpha) = 1, V(\alpha) \in [s_1,s_2))\\
&= \gamma^{-1}\{\mathbf{P}(Y_1 \in dt_1, Y_2 > t_2, \delta_1 = 1, V(\alpha) \in [s_1,s_2)) + \mathbf{P}(Y_1 > t_1, Y_2 \in dt_2, \delta_2 = 1, V(\alpha) \in [s_1,s_2))\\
&\quad + \mathbf{P}(Y_1 \in dt_1, Y_2 \in dt_2, \delta_1 = 1, \delta_2 = 0, V(\alpha) \in [s_1,s_2)) + \mathbf{P}(Y_1 \in dt_1, Y_2 \in dt_2, \delta_1 = 0, \delta_2 = 1, V(\alpha) \in [s_1,s_2))\\
&\quad + \mathbf{P}(Y_1 \in dt_1, Y_2 \in dt_2, \delta_1 = \delta_2 = 1, V(\alpha) \in [s_1,s_2))\}\\
&= \gamma^{-1}\mathbf{P}(C_1 \geq t_1, C_2 \geq t_2, V(\alpha) \in [s_1,s_2))\{\mathbf{P}(T_1 \in dt_1, T_2 > t_2) + \mathbf{P}(T_1 > t_1, T_2 \in dt_2) + \mathbf{P}(T_1 \in dt_1, T_2 \in dt_2)\}.
\end{aligned}$$

Thus equation (3.36) is true.

(b) For set A of the form $\{\tilde{Z}_i(\alpha) \geq s_3, V_i(\alpha) \geq s_1\}$ for any $0 \leq s_1, s_3 < z$, equation (3.36) can be proved similarly.

(c) For set A of the form $\{\tilde{Z}_i(\alpha) \in [s_3, s_4), \Delta_i(\alpha) = k, V_i(\alpha) \in [s_1, s_2)\}, k = 0, 1$ for any $0 \leq s_3 < s_4 \leq z$, equation (3.36) is also true since both sides are 0.

(d) Set A of other forms can be represented as union, intersection, complementary of the sets A in the above items **(a)**, **(b)** and **(c)**. Thus equation (3.36) holds.

■

The following definitions and lemma are needed to prove the large sample properties for the bivariate estimator $\hat{S}(z;\alpha)$ given in (3.31).

Definition 3.1.

1. Equation (3.36) implies that for fixed α, $M_i(ds;\alpha)$ and $M(ds;\alpha)$ are martingales with respect to the filtration $\mathscr{F}^i_{z;\alpha}$ and $\mathscr{F}_{z;\alpha} = \vee_{i=1}^n \mathscr{F}^i_{z;\alpha}$, respectively.
2. Let $H(t_1,t_2) := E[H_i(t_1,t_2)]$ and $H(z;\alpha) := E[H_i(z;\alpha)]$. Hence

$$\begin{aligned}
H(z;\alpha) &= H(t_1,t_2)\\
&= \mathbf{P}(Y_{1i} > t_1 \geq L_{1i}, Y_{2i} > t_2 \geq L_{2i})\\
&= \mathbf{P}(Y_1 > t_1 \geq L_1, Y_2 > t_2 \geq L_2 | T_1 > L_1, T_2 > L_2)\\
&= \mathbf{P}\big(\min\{T_1, C_1\} > t_1 \geq L_1, \min\{T_2, C_2\} > t_2 \geq L_2 | T_1 > L_1, T_2 > L_2\big)\\
&= \gamma^{-1} S(t_1,t_2) G(t_1,t_2)
\end{aligned} \tag{3.38}$$

where G is defined in (3.20).

Assumption 3.5.4 $\sup_\alpha \int_{[0,\tau_\alpha]}[S(s-;\alpha)G(s-;\alpha)]^{-1}\Lambda(ds;\alpha) < \infty$, *where* $\tau_\alpha = \sup_z\{z = \sqrt{t_1^2+t_2^2} : t_2/t_1 = \alpha, (t_1,t_2) \in \mathscr{A}\}$ *and* $\mathscr{A}$ *is defined in Assumption 3.5.3.*

Lemma 3.4. *Under assumptions 3.5.1, 3.5.2, 3.5.3 and 3.5.4, we have that when* $n \to \infty$, *for any* $\kappa \in (0,1)$,

$$n^{1-\kappa} \sup_{\alpha\in[0,\infty]} E\left[\int_{[0,\tau_\alpha]} I[H_{(n)}(s-;\alpha) = 0]\Lambda(ds;\alpha)\right]^2 \to 0. \tag{3.39}$$

Proof. The lemma can be proved similarly as Lemma 2.2. ∎

With the above lemma, we have the following theorems which imply the large sample properties of $\hat{\Lambda}(z;\alpha)$ and $\hat{S}(z;\alpha)$. Their proofs are omitted.

Theorem 3.1. *Under assumptions 3.5.1, 3.5.2, 3.5.3 and 3.5.4, for any* $(t_1,t_2) \in \mathscr{A}$ *and* $\kappa \in (0,1)$, *we have (consistency)*

$$\sup_{\alpha\in[0,\infty],z\in[0,\tau_\alpha]} E\left[\hat{\Lambda}(z;\alpha) - \Lambda(z;\alpha)\right]^2 \le O(1/n^{1-\kappa}) \tag{3.40}$$

and (asymptotic normality)

$$\sqrt{n}\left[\hat{\Lambda}(z;\alpha) - \Lambda(z;\alpha)\right] \Rightarrow \xi(z;\alpha) \tag{3.41}$$

where $\xi(z;\alpha)$ *is a zero-mean Gaussian process with independent increments and variance function*

$$\sigma_\Lambda^2(z;\alpha) = \int_{[0,z]} [H(s;\alpha)]^{-1}\Lambda(ds;\alpha).$$

∎

Theorem 3.2. *Under assumptions 3.5.1, 3.5.2, 3.5.3 and 3.5.4, for any* $(t_1,t_2) \in \mathscr{A}$ *and* $\kappa \in (0,1)$, *we have (consistency)*

$$\sup_{\alpha\in[0,\infty],z\in[0,\tau_\alpha]} E\left[\hat{S}(z;\alpha) - S(z;\alpha)\right]^2 \le O(1/n^{1-\kappa}) \tag{3.42}$$

and (asymptotic normality)

$$\sqrt{n}\left[\hat{S}(z;\alpha) - S(z;\alpha)\right] \Rightarrow \eta(t;\alpha) \tag{3.43}$$

where $\eta(t;\alpha)$ *is a zero-mean Gaussian process with independent increments and variance function*

$$\sigma_S^2(z;\alpha) = S^2(z;\alpha)\int_{[0,z]} [H(s;\alpha)]^{-1}\Lambda(ds;\alpha). \tag{3.44}$$

∎

Substituting the consistent estimators $\hat{\Lambda}(z;\alpha)$ and $\hat{S}(z;\alpha)$ into (3.44), we can obtain a consistent estimator for $\sigma^2(z;\alpha)$ which is

$$\hat{\sigma}_S^2(z;\alpha) = \hat{S}^2(z;\alpha)\int_{[0,z]} \frac{N(ds;\alpha)}{[H_{(n)}(s-;\alpha) - \Delta N(s;\alpha)]H_{(n)}(s-;\alpha)}. \tag{3.45}$$

The estimator $\hat{S}(t_1,t_2) = \hat{S}(z;\alpha)$ has nice large sample properties as shown above. The monotonicity of $S(t_1,t_2)$ means that $S(t_1,t_2) \ge S(v_1,v_2)$ for $\boldsymbol{t} = (t_1,t_2) < \boldsymbol{v} = (v_1,v_2)$. The survival function estimator $\hat{S}(t_1,t_2)$ and the distribution function estimator $1 - \hat{S}(0,t_2) - \hat{S}(t_1,0) + \hat{S}(t_1,t_2)$ however may not be monotone as $\hat{S}(t_1,t_2)$ may be less than $\hat{S}(v_1,v_2)$ for $\boldsymbol{t} < \boldsymbol{v}$ with $v_2/v_1 \ne t_2/t_1$. The reason is that the estimator is based on the transformed data in the polar coordinate system. In the following subsection, we introduce an IPW estimator $\hat{F}(t_1,t_2)$ for the joint distribution function based on the estimator $\hat{S}(t_1,t_2)$. The IPW estimator $\hat{F}(t_1,t_2)$ is monotone and gives non-negative probability mass on each point.

Note that the above estimates can also be amended for small sample sizes, similarly as (2.42) and (2.43). Details are omitted here.

3.5.5 The graphical interpretation of the transformation estimator

When estimating the bivariate survival function $S(t_1,t_2),(t_1,t_2) \in \mathbf{R}^+ \times \mathbf{R}^+$, one may fix t_2 and estimate $S(t_1,t_2)$ for $t_1 \in \mathbf{R}^+$ first; then let t_2 varies in $\mathbf{R}^+$. Such a naive approach treat the whole estimation procedure as two steps of univariate survival function estimation, i.e. estimation of $S(t_1|T_2 > t_2)$ and $S(t_2)$. However, such naive approach will not give good estimates since we are considering nonparametric estimation and the conditional distribution $S(t_1|T_2 > t_2)$ may not be estimated well.

The proposed transformation method changes the estimation of $\{S(t_1,t_2),(t_1,t_2) \in \mathbf{R}^+ \times \mathbf{R}^+\}$ to $\{S(z;\alpha),(z,\alpha) \in \mathbf{R}^+ \times \mathbf{R}^+\}$, which can also be estimated via two steps: estimation of $\{S(z;\alpha), z \in \mathbf{R}^+\}$ with α fixed and then let α varies from 0 to ∞. The advantage of using such transformation is that when we estimate $S(z;\alpha)$ for fixed α, it is based on the transformed data $Z(\alpha)$ (including information of both T_1 and T_2). Given a data set of size n, when estimating $\{S(z;\alpha), z \in \mathbf{R}^+\}$ for each value of α, we will still have n transformed observations. Therefore, when we change the value of α, we will not lose any information by doing the transformation.

The transformed data can be viewed as the minimum of the two distances, respectively, from $(0,0)$ to the points of the vertical projection and horizontal projection of (T_1,T_2) on the line through $(0,0)$ and having gradient α. This is shown in Figure 3.1. Therefore the transformation data can be viewed as observations (univariate) on the line through $(0,0)$ and having gradient α. The estimation of $\{S(z;\alpha), z \in \mathbf{R}^+\}$ is actually univariate survival function estimation.

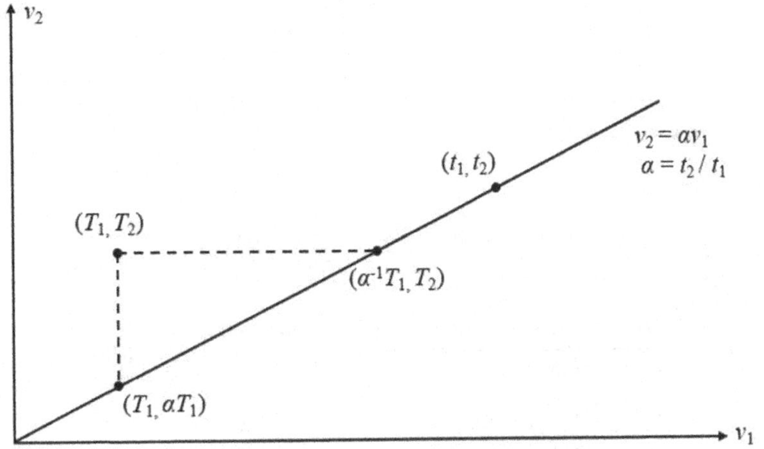

Fig. 3.1: Data transformation.

3.5.6 The IPW bivariate distribution estimator

Under the assumptions described at the beginning of this section, consider the sub-distribution function

$$
\begin{aligned}
F^*(t_1,t_2) &= \mathbf{P}(Y_{1i} \le t_1, Y_{2i} \le t_2, \delta_{1i} = 1, \delta_{2i} = 1) \\
&= \mathbf{P}(T_1 \le t_1, T_2 \le t_2, T_1 \le C_1, T_2 \le C_2 | L_1 < T_1, L_2 < T_2) \\
&= \gamma^{-1}\mathbf{P}(L_1 < T_1 \le C_1, L_2 < T_1 \le C_2, T_1 \le t_1, T_2 \le t_2) \\
&= \gamma^{-1} \int_{[0,t_2]} \int_{[0,t_1]} G(v_1-, v_2-) F(dv_1, dv_2)
\end{aligned}
$$

where G is defined in (3.20).

Then following (3.38), we immediately have that

$$F(t_1,t_2) = \int_{[0,t_2]}\int_{[0,t_1]} \frac{S(v_1-,v_2-)}{H(v_1-,v_2-)} F^*(dv_1,dv_2). \tag{3.46}$$

Therefore an IPW estimator for $F(t_1,t_2)$ is given by

$$\hat{F}(t_1,t_2) = \int_{[0,t_2]}\int_{[0,t_1]} I[H_{(n)}(v_1-,v_2-)>0]\frac{\hat{S}(v_1-,v_2-)}{H_{(n)}(v_1-,v_2-)} F^*_{(n)}(dv_1,dv_2), \tag{3.47}$$

where $F^*_{(n)}(v_1,v_2) = n^{-1}\sum_{i=1}^{n} I[Y_{1i}\le t_1, Y_{2i}\le t_2, \delta_{1i}=1, \delta_{2i}=1]$. Equivalently, the IPW estimator in (3.47) can also be written as

$$\hat{F}(t_1,t_2) = \int_{[0,t_2]}\int_{[0,t_1]} I[H_{(n)}(v_1-,v_2-)>0]\frac{\hat{S}(s-;\beta)}{H_{(n)}(v_1-,v_2-)} F^*_{(n)}(dv_1,dv_2), \tag{3.48}$$

where

$$\begin{aligned} s &= \sqrt{v_1^2+v_2^2} \\ \beta &= v_2/v_1. \end{aligned} \tag{3.49}$$

To prove consistency and asymptotic normality, we rewrite the bivariate distribution estimator $\hat{F}(t_1,t_2)$ as sum of i.i.d random variables. We have

$$\begin{aligned}
&\sqrt{n}\left[\hat{F}(t_1,t_2) - F(t_1,t_2)\right] \\
&= \sqrt{n}\int_{[0,t_2]}\int_{[0,t_1]} \left[\frac{I[H_{(n)}(v_1-,v_2-)>0]\hat{S}(s-;\beta)}{H_{(n)}(v_1-,v_2-)} - \frac{S(s-;\beta)}{H(v_1-,v_2-)}\right]\left[F^*_{(n)}(dv_1,dv_2) - F^*(dv_1,dv_2)\right] \\
&+ \sqrt{n}\int_{[0,t_2]}\int_{[0,t_1]} \left[\frac{I[H_{(n)}(v_1-,v_2-)>0]\hat{S}(s-;\beta)}{H_{(n)}(v_1-,v_2-)} - \frac{S(s-;\beta)}{H(v_1-,v_2-)}\right] F^*(dv_1,dv_2) \\
&+ \sqrt{n}\int_{[0,t_2]}\int_{[0,t_1]} \frac{S(s-;\beta)}{H(v_1-,v_2-)}\left[F^*_{(n)}(dv_1,dv_2) - F^*(dv_1,dv_2)\right].
\end{aligned} \tag{3.50}$$

Lemma 3.5. *The first term at the right hand side of* (3.50)

$$\sqrt{n}\int_{[0,t_2]}\int_{[0,t_1]} \left[\frac{I[H_{(n)}(v_1-,v_2-)>0]\hat{S}(s-;\beta)}{H_{(n)}(v_1-,v_2-)} - \frac{S(s-;\beta)}{H(v_1-,v_2-)}\right]\left[F^*_{(n)}(dv_1,dv_2) - F^*(dv_1,dv_2)\right]$$

converges to 0 in probability as $n \to \infty$.

The proof of this lemma can be found in the Appendix A.7 in Dai and Fu (2012).

Using Lemma 3.5 we have

$$
\begin{aligned}
&\sqrt{n}\big[\hat{F}(t_1,t_2)-F(t_1,t_2)\big]\\
&=\sqrt{n}\int_{[0,t_2]}\int_{[0,t_1]}\left[\frac{I[H_{(n)}(v_1-,v_2-)>0]\hat{S}(s-;\beta)}{H_{(n)}(v_1-,v_2-)}-\frac{S(s-;\beta)}{H(v_1-,v_2-)}\right]F^*(dv_1,dv_2)\\
&\quad+\sqrt{n}\int_{[0,t_2]}\int_{[0,t_1]}\frac{S(s-;\beta)}{H(v_1-,v_2-)}\big[F^*_{(n)}(dv_1,dv_2)-F^*(dv_1,dv_2)\big]+o_p(1)\\
&:=I+II+o_p(1)
\end{aligned}
\tag{3.51}
$$

where term I can be written as

$$
\begin{aligned}
I=&\sqrt{n}\int_{[0,t_2]}\int_{[0,t_1]}S(s-;\beta)\left[\frac{H(v_1-,v_2-)-H_{(n)}(v_1-,v_2-)}{H^2(v_1-,v_2-)}\right]F^*(dv_1,dv_2)\\
&+\sqrt{n}\int_{[0,t_2]}\int_{[0,t_1]}\frac{1}{H(v_1-,v_2-)}\big[\hat{S}(v_1-,v_2-)-S(v_1-,v_2-)\big]F^*(dv_1,dv_2)\\
&+\sqrt{n}\int_{[0,t_2]}\int_{[0,t_1]}\left[\frac{I[H_{(n)}(v_1-,v_2-)>0]\hat{S}(s-;\beta)}{H_{(n)}(v_1-,v_2-)}-\frac{S(s-;\beta)}{H(v_1-,v_2-)}\right]\left[\frac{H(v_1-,v_2-)-H_{(n)}(v_1-,v_2-)}{H(v_1-,v_2-)}\right]F^*(dv_1,dv_2)\\
&+\sqrt{n}\int_{[0,t_2]}\int_{[0,t_1]}\frac{\hat{S}(s-;\beta)}{H(v_1-,v_2-)}\big(I[H_{(n)}(v_1-,v_2-)>0]-1\big)F^*(dv_1,dv_2)\\
:=&I_1+I_2+I_3+I_4.
\end{aligned}
\tag{3.52}
$$

Now we consider I_2. From Theorem 3.3 in Dai and Fu (2012), we know that

$$
-\sqrt{n}S(v_1-,v_2-)\int_{[0,s]}\frac{1}{H(b-;\beta)}M(db;\beta)+r_n(s;\beta)=\sqrt{n}\big[\hat{S}(v_1-,v_2-)-S(v_1-,v_2-)\big],
$$

where $\sup_{s,\beta}E[r_n(s;\beta)]^2\to 0$. If we let $M_i'(s;\beta)=\int_{[0,s]}\frac{1}{H(b-;\beta)}M_i(db;\beta)$, then the term I_2 can be written as

$$
I_2=-\sqrt{n}\sum_{i=1}^{n}\int_{[0,t_2]}\int_{[0,t_1]}M_i'(s;\beta)F(dv_1,dv_2)+o_p(1),
$$

which means that I_2 can be represented as sum of i.i.d. random variables, asymptotically. On the other hand, I_1 is also a sum of i.i.d random variables. The terms I_3 and I_4 are both $o_p(1)$.

In summary we have

$$
\begin{aligned}
&\sqrt{n}\big[\hat{F}(t_1,t_2)-F(t_1,t_2)\big]\\
&=\sqrt{n}\int_{[0,t_2]}\int_{[0,t_1]}\left[\frac{H(v_1-,v_2-)-H_{(n)}(v_1-,v_2-)}{H(v_1-,v_2-)}\right]F(dv_1,dv_2)\\
&\quad-\sqrt{n}\sum_{i=1}^{n}\int_{[0,t_2]}\int_{[0,t_1]}M_i'(s;\beta)F(dv_1,dv_2)\\
&\quad+\sqrt{n}\int_{[0,t_2]}\int_{[0,t_1]}\frac{S(s-;\beta)}{H(v_1-,v_2-)}\big[F^*_{(n)}(dv_1,dv_2)-F^*(dv_1,dv_2)\big]+o_p(1)
\end{aligned}
\tag{3.53}
$$

which is a sum of i.i.d. terms. Therefore we have the large sample properties of $\hat{F}(t_1,t_2)$ given in the following theorem.

Theorem 3.3. *Let* $\alpha_i=Y_{2i}^*/Y_{1i}^*$ *and*

$$\eta_i(t_1,t_2) = I\left[Y_{1i}^* \le t_1, Y_{2i}^* \le t_2, \delta_{1i}^* = 1, \delta_{2i}^* = 1\right] \frac{S(\tilde{Z}_i(\alpha_i)-;\alpha_i)}{H(\tilde{Z}_i(\alpha_i)-;\alpha_i)} - F(t_1,t_2),$$

$$\mu_i(t_1,t_2) = \int_{[0,t_2]}\int_{[0,t_1]} M_i'(s;\alpha)F(dv_1,dv_2),$$

$$\xi_i(t_1,t_2) = \int_{[0,t_2]}\int_{[0,t_1]} \frac{H(v_1-,v_2-)-H_i(v_1-,v_2-)}{H(v_1-,v_2-)}F(dv_1,dv_2).$$

For any (t_1,t_2) in $\mathscr{A}$, we have

$$\sqrt{n}\left[\hat{F}(t_1,t_2) - F(t_1,t_2)\right] \Rightarrow N(0,\sigma^2(t_1,t_2)),$$

where

$$\sigma^2(t_1,t_2) = Var\left[\xi_i(t_1,t_2) + \eta_i(t_1,t_2) - \mu_i(t_1,t_2)\right].$$

Theorem 3.3 gives the asymptotic normality of the bivariate distribution estimator $\hat{F}(t_1,t_2)$. The proof can be found in Dai and Fu (2012). We can calculate the estimates $\hat{\xi}_i$, $\hat{\eta}_i$ and $\hat{\mu}_i$, by replacing S, F and H with their consistent estimators in the formulas of ξ_i, η_i and μ_i. Then an estimate of $\sigma^2(t_1,t_2)$ is given by the sample variance of $\hat{\xi}_i$, $\hat{\eta}_i$ and $\hat{\mu}_i$.

3.6 Example

We now apply the method discussed in this chapter to analyze the Edinburgh hepatitis C data in Fu et al. (2007). The cohort consists of 387 hepatitis C virus (HCV) infected patients who were referred to the Edinburgh Royal Infirmary liver clinic by the end of 1999. Sixty-three of them were observed to have been progressed to liver cirrhosis before their last biopsy diagnosis time. The two time durations of interest are the incubation period from HCV infection to cirrhosis denoted by T_2 and the time period from disease infection to the referral to the liver clinic denoted by T_1. The time T_2 is subject to right censoring at the last diagnosis follow-up time C. We only observe $Y = \min\{T_2, C\}$ and the censoring indicator $\delta = I[T_2 \le C]$. An individual information is only available if the patient was referred to the clinic cohort before the end of 1999, that is, the data are subject to a univariate truncation $T_1 \le D$, where D is the time period from disease infection to the end of 1999. Here T_1 is right truncated at D. We can use the negative values of T_1 and D to obtain left-truncated observations. All the variables are measured in years.

The plot of the estimated joint distribution function is shown in Figure 3.2. Particularly we are interested in the estimated progression rate to cirrhosis, explained by the estimated marginal distribution function of T_2 in Figure 3.3. The estimated percentage of patients with cirrhosis is 2% after 20 years of HCV infection and 20% after 30 years based on our proposed method. Comparing the bivariate and univariate analyses, as shown in Figure 3.3, the bivariate analysis shows much lower estimated progression rates to cirrhosis. The estimate of progression rate from the univariate analysis, without taking account of the referral bias, is 9% within 20 years and 42% within 30 years, which are seriously overestimated (Fu et al., 2007). According to the published reports based on community cohort studies in difference countries (for a summary see Freeman et al., 2001), the 20-year progression rate is 2-7% and the 30-year progression rate is 5-25% among all the HCV-infected patients in the community. The truncation probability estimate is 0.081, which means that only about 8% of the HCV-infected population in the Edinburgh area were included in the clinical cohort due to the truncation at the end of 1999.

3.7 Discussion

Nonparametric analysis is usually preferable in survival analysis, especially in bivariate survival analysis, since most parametric distribution cannot satisfy the requirements of censoring and truncation distri-

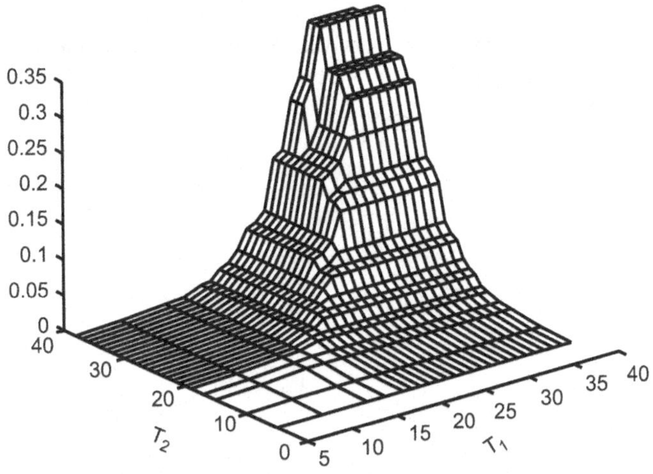

Fig. 3.2: The joint distribution for T_1 and T_2

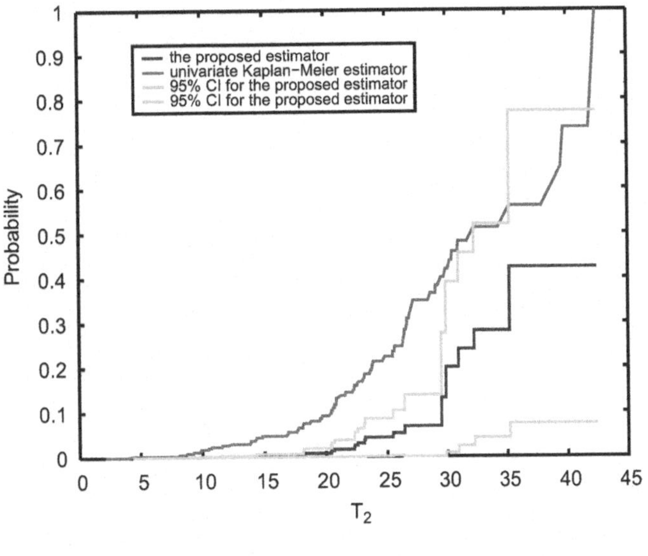

Fig. 3.3: The marginal distribution for T_2

butions or the requirements of dependence between variables. One may refer to Hougaard (2000) for a detailed introduction of the dependence structure between bivariate survival times. In some bivariate (multivariate) survival analysis, each subject may experience two (or more) event times and Markov transition model may be used. The Markov assumption, however, may not be realistic in practice. The nonparametric bivariate survival function estimator could be used in those scenarios to release the Markov assumption.

Bivariate survival analysis is much more challenging than the univariate case, especially under censoring and truncation. One challenge is that difficulties arise in the martingale theory on the plane (Merzbach and Nualart, 1988) because we do not have a complete ordering on $\mathbf{R}_+^2$. If we consider the approach in Dabrowska (1988), although we can write the bivariate survival function as a product integral as well

(3.5), the resulting estimator may not be a monotone function. The transformation estimator is not monotone either. Nevertheless, non-monotone estimators can be transformed to monotone estimators via the IPW method.

Including explanatory variables in bivariate survival analysis leads to a more challenging field, which will be discussed in the following chapters.

Chapter 4
Accelerated failure time model for truncated and censored survival data

4.1 Introduction

In this chapter, we consider the accelerated failure time (AFT) model (Cox and Oakes, 1984; Kalbfleisch and Prentice, 2002) to study the effect of the covariates on the time to event. The AFT model relates the event time to the covariates through the following relationship

$$\log T = \boldsymbol{W}\beta + \varepsilon, \tag{4.1}$$

where T is the actual event time, $\boldsymbol{W}$ is a random vector of covariates and β is an unknown parameter to be estimated. We may assume that ε follows a particular parametric distribution, for example, normal distribution. However, it may not be easy to justify the distribution model for ε in practice, as it could be a heavy tail distribution or even a mixture of distributions due to the complexity of the data. Therefore it is more reliable to choose an unspecified distribution function F_ε, i.e. a nonparametric model for ε. In addition, it is assumed that ε and $\boldsymbol{W}$ are uncorrelated. For notational convenience, we can denote T as the logarithm of the actual event time. Then the model (4.1) can be studied as a standard linear regression model

$$T = \boldsymbol{W}\beta + \varepsilon, \tag{4.2}$$

since the effect of the covariates $\boldsymbol{W}$ is linearly related to the response T. Therefore, the AFT model is easier to interpret, more relevant to clinicians, and can be a useful alternative to the commonly used Cox proportional hazards model.

4.1.1 Censoring only

When the event time is subject to only random censoring, model (4.2) has been well studied. Miller (1976) extended the analysis for a standard linear regression model to the case that data may be randomly right censored. His regression parameter estimator is weighted linear combinations of the uncensored observations where the weights are derived from the well-known Kaplan-Meier (KM) estimator of F_ε. For the estimate $\hat{\beta}$ to be asymptotic consistent, it requires that as the covariates $\boldsymbol{W}$ changes, the censoring distributions shift along the same line as the distributions of the event time (Miller and Harplen, 1982). However, this condition is much stronger than is necessary and will be rarely satisfied in practice.

Buckley and James (1979) proposed an estimation method which does not require such restrictions. The main difference between their method and that of Miller (1976) is that the data are modified rather than the residual sum of squares. In the absence of censoring, the classic least squares estimator for the regression parameter β in model (4.2) is obtained by minimizing

Analysis for Time-to-Event Data under Censoring and Truncation.
http://dx.doi.org/10.1016/B978-0-12-805480-2.50004-6,

$$n^{-1}\sum_{i=1}^{n}(T_i - W_i\beta)^2 \tag{4.3}$$

with respect to β. In the presence of right censoring, the observed data are $\{(X_i, W_i, \delta_i);\ i = 1,2,\ldots,n\}$, where $X_i = \min(T_i, C_i)$ and $\delta_i = I[T_i \le C_i]$. The values of T_i associated with $\delta_i = 0$ are unknown, so that (4.3) cannot be utilized directly to estimate β. Therefore, Buckley and James (1979) modified the above procedure by substituting the observed X_i by

$$X_i^* = \delta_i X_i + (1-\delta_i)E[T_i | T_i > X_i],\ i = 1,\ldots,n. \tag{4.4}$$

In other words, the event times for all the censored observations are replaced by their conditional expectation $E[T_i|T_i > X_i]$. Since $E[T_i|T_i > X_i]$ is unknown, they suggested using a self-consistency approach to estimate it from the KM estimator of $F_{\tilde{\varepsilon}}$, which is the distribution of $\tilde{\varepsilon}_i = X_i - W_i\beta,\ i = 1,\ldots,n$. Then the replacing variable X_i^* in (4.4) can be estimated by

$$\hat{X}_i^*(\beta) = \delta_i X_i + (1-\delta_i)\left[W_i\beta + \frac{\sum_{j:\tilde{\varepsilon}_j > \tilde{\varepsilon}_i}\omega_j^*(\beta)\tilde{\varepsilon}_j}{1-\hat{F}_{\tilde{\varepsilon}}(\tilde{\varepsilon}_i)}\right],\ j = 1,\ldots,n, \tag{4.5}$$

where

$$\omega_j^*(\beta) = \frac{\omega_j(\beta)}{\sum_{k=1}^{n}\omega_k(\beta)},\quad \omega_k(\beta) = d\hat{F}_{\tilde{\varepsilon}}(\tilde{\varepsilon}_k). \tag{4.6}$$

Then the estimated $\hat{X}_i^* = \hat{X}_i^*(\beta)$ can be used to estimate β by minimising

$$n^{-1}\sum_{i=1}^{n}(\hat{X}_i^* - W_i\beta)^2. \tag{4.7}$$

The above arguments actually give an iteration algorithm for the estimation of β. By treating all the observations as uncensored, an iterative estimation procedure was proposed with the initial estimator

$$\hat{\beta}^{(0)} = \{(W-\bar{W})^T(W-\bar{W})\}^{-1}(W-\bar{W})^T X, \tag{4.8}$$

and the estimator at the $k+1$ step

$$\hat{\beta}^{(k+1)} = \{(W-\bar{W})^T(W-\bar{W})\}^{-1}(W-\bar{W})^T \hat{X}^*(\hat{\beta}^{(k)}), \tag{4.9}$$

where

$$X = (X_1,\ldots,X_n)^T,\quad \hat{X}^*(\hat{\beta}^{(k)}) = (\hat{X}_1^*(\hat{\beta}^{(k)}),\ldots,\hat{X}_n^*(\hat{\beta}^{(k)}))^T. \tag{4.10}$$

Here if the regression parameter β is a p-dimensional vector, then W is a $n \times p$ covariates matrix with elements W_{ij}, and $\bar{W}$ is a $n \times p$ matrix with elements $n^{-1}\sum_{i=1}^{n} W_{ij}$, $i = 1,\ldots,n$, $j = 1,\ldots,p$. The limit of the sequence in (4.9) is the estimate of β. However, the iterations may also settle down to oscillation between two values as that for Miller's estimator, although the values are closer according to Buckley and James (1979). If so, the average of the two values can be used as the estimate of β. Their method only assumes that the event time T and the censoring time C are conditionally independent, given the covariates W, which also needs to be satisfied in Miller's approach. The inconsistency problems in Miller's approach appear to be overcome since the strict assumption of the censoring distribution is not necessary here, which means the BJ estimator does not depend on particular censoring patterns. The asymptotic properties of the BJ estimator were studied by Ritov (1990) and Lai and Ying (1991b). They showed that, with a slight modification to the tail, any consistent root of the estimating equation proposed by Buckley and James must be asymptotically normal and that the estimator is semi-parametrically efficient when the underlying error distribution is normal, which is also a well-known property of the least squares estimator for uncensored data. However, the limiting covariance matrix is difficult to estimate directly since it involves the unknown hazard function of the error term.

The approaches proposed by Miller (1976) and Buckley and James (1979) for the AFT model assumed that the event time T and the censoring time C are independent conditionally on the covariates $\boldsymbol{W}$. Under this assumption, the distribution of C depends on $\boldsymbol{W}$, which further implies that the distribution of the error term, F_ε, depends on the unknown parameter $\boldsymbol{\beta}$. Therefore, iteration algorithms are required to locate the estimate of $\boldsymbol{\beta}$.

There are some other existing works on the AFT model which assume a stronger assumption that T and C are independent. Under this independence assumption, Koul et al. (1981) constructed the synthetic data

$$X_i^*(G) = \frac{\delta_i}{1-G(X_i-)}X_i,\ i=1,\ldots,n, \tag{4.11}$$

where $G(t) = \mathbf{P}(C \le t)$. Substituting G by its KM estimator $\hat{G}$, the estimate of $\boldsymbol{\beta}$ can be obtained by

$$\hat{\boldsymbol{\beta}} = \left\{(\boldsymbol{W}-\bar{\boldsymbol{W}})^T(\boldsymbol{W}-\bar{\boldsymbol{W}})\right\}^{-1}(\boldsymbol{W}-\bar{\boldsymbol{W}})^T\hat{\boldsymbol{X}}^*(\hat{G}), \tag{4.12}$$

where $\hat{\boldsymbol{X}}^*(\hat{G}) = \left(\hat{X}_1^*(\hat{G}),\ldots,\hat{X}_n^*(\hat{G})\right)$, $\boldsymbol{W}$ and $\bar{\boldsymbol{W}}$ follow the same definitions as above. The independence of T and C was also assumed for the weighted least squares (WLS) estimator proposed by He and Wong (2003). Their estimate of $\boldsymbol{\beta}$ is a vector that minimizes

$$\sum_{i=1}^{n}\frac{\delta_i}{1-\hat{G}(X_i-)}(X_i-\boldsymbol{W}_i\boldsymbol{\beta})^2, \tag{4.13}$$

and hence is the same as the estimator proposed by Stute (1993). As pointed out in a comparison study (Bao et al., 2007), when the censoring mechanism depends on covariates, the false results from the KSV estimator may be avoided by the use of the WLS estimator, since the WLS approach incorporates both weighted responses and weighted covariates.

4.1.2 In the presence of truncation

Under random truncation, the observation of $(T,\boldsymbol{W})$ is interfered by another independent random variable L such that $(T,\boldsymbol{W},L)$ can be observed only if $T > L$ (left truncation) or $T < L$ (right truncation). In either case, some subjects may be missing and only a biased sample $\left\{(T_i,\boldsymbol{W}_i,L_i);\ i=1,2,\ldots,n\right\}$ can be observed. If the data are complete, we can estimate the regression parameter $\boldsymbol{\beta}$ and the error distribution F_ε using the ordinary least squares (OLS) procedure, which is based on the relationship $E[T|\boldsymbol{W}] = \boldsymbol{W}\boldsymbol{\beta}$. However, if the data are truncated, the conditional expectation $E[T|\boldsymbol{W},T>L]$ or $E[T|\boldsymbol{W},T<L]$ is no longer a linear function of $\boldsymbol{\beta}$. The OLS procedure cannot produce correct estimates for $\boldsymbol{\beta}$ and F_ε.

P. K. Bhattacharya and Yang (1983) addressed the estimation of the regression parameter in a simple linear regression model, while the truncation variable L was assumed to be a known constant. The Kendall rank correlation coefficient (Kendall's τ) and weighted medians were used to estimate the regression parameter $\boldsymbol{\beta}$. However, their iterative algorithm requires a simultaneous estimation of F_ε and $\boldsymbol{\beta}$, which is computationally intensive.

Lai and Ying (1991c) developed rank regression methods in the presence of both left truncation and right censoring. They considered the case that an individual can be observed only if $T > L$ and T itself may be also subject to right censoring by another random variable C. The observed left-truncated and right-censored (LTRC) data was denoted by $(X_i,L_i,\delta_i,\boldsymbol{W}_i)$, $i=1,\ldots,n$, where $X_i = \min(T_i,C_i)$ and $\delta_i = I[T_i \le C_i]$. Define the residuals $\tilde{\varepsilon}_i(\boldsymbol{\beta}) = X_i - \boldsymbol{W}_i\boldsymbol{\beta}$ and let $\tilde{\varepsilon}_{(1)}(\boldsymbol{\beta}) \le \cdots \le \tilde{\varepsilon}_{(k)}(\boldsymbol{\beta})$ denote all the ordered uncensored residuals. For $i=1,\ldots,k$, they constructed the at-risk sets

$$\mathscr{I}(i,\boldsymbol{\beta}) = \left\{j \le n : L_j - \boldsymbol{W}_j\boldsymbol{\beta} \le \tilde{\varepsilon}_{(i)}(\boldsymbol{\beta}) \le X_j - \boldsymbol{W}_j\boldsymbol{\beta}\right\}, \tag{4.14}$$

and proposed the following estimating equation for $\boldsymbol{\beta}$ based on the at-risk sets defined in (4.14),

$$U(\beta) = \sum_{i=1}^{n} \left\{ \mathbf{W}_i - \frac{\sum_{j=1}^{n} \mathbf{W}_j I[j \in \mathscr{J}(i,\beta)]}{\sum_{j=1}^{n} I[j \in \mathscr{J}(i,\beta)]} \right\} = \mathbf{0}. \tag{4.15}$$

Lai and Ying (1994) also proposed a general missing information principle and constructed a weighted M-estimator of regression parameter under the AFT model (4.2) for LTRC data. When $\{(T_i, \mathbf{W}_i);\ i = 1,2,\ldots,n\}$ can be fully observed, the classical M-estimator (Huber, 1973) is defined by the estimating equation

$$\sum_{i=1}^{n} \mathbf{W}_i \rho'(T_i - \mathbf{W}_i\beta) = 0, \tag{4.16}$$

where the function ρ is a differentiable loss function. In particular, when $\rho(u) = u^2$, $\hat{\beta}$ from (4.16) reduces to the OLS estimate, and it reduces to the maximum likelihood estimate of β when $\rho(u) = -\log F_\varepsilon(u)$. In the presence of left truncation and right censoring on T, for $i = 1,\ldots,n$, let $\tilde{\varepsilon}_i(\beta) = X_i - \mathbf{W}_i\beta$, $\eta_i(\beta) = L_i - \mathbf{W}_i\beta$, and let $F_{\tilde{\varepsilon}}$ denote the distribution function of $\tilde{\varepsilon}_i(\beta)$. Lai and Ying (1994) used similar idea as Buckley and James (1979) and defined the synthetic data as

$$\hat{X}_i^*(\beta) = \delta_i \rho'\left(\tilde{\varepsilon}_i(\beta)\right) + (1-\delta_i)\frac{\int_{[\tilde{\varepsilon}_i(\beta),\infty]} \rho'(u) dF_{\tilde{\varepsilon}}(u)}{1 - F_{\tilde{\varepsilon}}\left(\tilde{\varepsilon}_i(\beta)\right)} - \frac{\int_{[\eta_i(\beta),\infty]} \rho'(u) dF_{\tilde{\varepsilon}}(u)}{1 - F_{\tilde{\varepsilon}}\left(\eta_i(\beta)\right)}. \tag{4.17}$$

Replacing $F_{\tilde{\varepsilon}}$ in (4.17) by the product-limit estimator $\hat{F}_{\tilde{\varepsilon}}$, the estimating equation for β is given by

$$\sum_{i=1}^{n} \mathbf{W}_i \left\{ \delta_i \rho'\left(\tilde{\varepsilon}_i(\beta)\right) + (1-\delta_i)\frac{\int_{[\tilde{\varepsilon}_i(\beta),\infty]} \rho'(u) d\hat{F}_{\tilde{\varepsilon}}(u)}{1 - \hat{F}_{\tilde{\varepsilon}}\left(\tilde{\varepsilon}_i(\beta)\right)} - \frac{\int_{[\eta_i(\beta),\infty]} \rho'(u) d\hat{F}_{\tilde{\varepsilon}}(u)}{1 - \hat{F}_{\tilde{\varepsilon}}\left(\eta_i(\beta)\right)} \right\} = \mathbf{0}. \tag{4.18}$$

Gross and Lai (1996) studied the AFT model in (4.2) from a view point of curve fitting or approximation theory and proposed another kind of M-estimator. They defined

$$\tau = \inf\{t : G(t) > 0\}, \quad \tau^* = \inf\{t > \tau : G(t)S(t) = 0\}, \tag{4.19}$$

where $G(t) = \mathbf{P}(L \le t \le C)$ and $S(t) = \mathbf{P}(T > t)$. Thus τ is the lower boundary of the support of the truncation distribution, and τ^* is the upper boundary of the support of the distribution of the observed $X = \min(T, C)$. Then only the conditional distribution $F(t|\tau) = \mathbf{P}(T \le t | T \ge \tau)$ can be nonparametrically estimated from the observed data, and only at times $t \le \tau^*$ (Keiding and Gill, 1990; Lai and Ying, 1991a; Gross and Lai, 1996). Their estimating equation for β is given by

$$n^{-1} \sum_{i=1}^{n} \delta_i I[X_i \in \mathscr{A}] \mathbf{W}_i \rho'(X_i - \mathbf{W}_i\beta)\hat{S}(X_i) = 0, \tag{4.20}$$

where $\mathscr{A}$ denotes some intervals such that $\mathscr{A} \in (\tau, \tau^*)$, ρ is a general loss function, and $\hat{S}$ is the product-limit estimator of $S(t) = \mathbf{P}(T > t)$. The estimating equation in (4.20) implies that the estimate of β should be the values giving the best fit in terms of a certain weighted M-estimation criterion. Therefore in their method, the β cannot be interpreted as the coefficient of an assumed linear regression model $E[T|\mathbf{W}] = \mathbf{W}\beta$, since their estimating equation only defines β as the coefficient of the best linear predictor $\mathbf{W}\beta$ of T in the sense of minimizing certain expected measure of the prediction error.

He and Yang (2003) considered the AFT model in (4.2) for left truncated data. Using similar idea for the censoring only case in He and Wong (2003), they constructed WLS estimator for β where the weights were random quantities determined by the product limit estimates of the distribution function of L. Their approach does not require iterative algorithm and can be extended to the case with LTRC data, which we will introduce in the next section.

4.2 WLS estimator for univariate LTRC data under AFT model

Consider the AFT model in (4.2), where the event time T may be subject to random left truncation by L and may be right censored by the censoring time C. The left truncation condition, $L < T$, is defined in Definition 2.1. The observed data are denoted by $(X_i, L_i, \delta_i, \boldsymbol{W}_i)$, $i = 1, \ldots, n$, where $X_i = \min(T_i, C_i)$ and $\delta_i = I[T_i \leq C_i]$. Let $S(t) = \mathbf{P}(T > t)$ be the survival function of T, and let $F(t, \boldsymbol{w}) = \mathbf{P}(T \leq t, \boldsymbol{W} \leq \boldsymbol{w})$ be the joint distribution function of $(T, \boldsymbol{W})$. Under Condition 2.1.8, we consider the sub-distribution function

$$\begin{aligned} F^*(t,\boldsymbol{w}) &= \mathbf{P}(X_i \leq t, \boldsymbol{W}_i \leq \boldsymbol{w}, \delta_i = 1) \\ &= \mathbf{P}(X \leq t, \boldsymbol{W} \leq \boldsymbol{w}, \delta = 1 | L < T) \\ &= \gamma^{-1}\mathbf{P}(L < T \leq C, T \leq t, \boldsymbol{W} \leq \boldsymbol{w}) \\ &= \gamma^{-1}\int_{[a,t]}\int_{\boldsymbol{v}\leq\boldsymbol{w}} \mathbf{P}(L < u \leq C) F(du, d\boldsymbol{v}), \end{aligned} \tag{4.21}$$

where $a = \inf\{t : \mathbf{P}(L < t \leq C) > 0\}$, and $\gamma = \mathbf{P}(T > L) > 0$ denotes the truncation probability, not depending on $\boldsymbol{W}$.

Define $H_i(t) = I[L_i < t \leq X_i]$ and $H(t) = E[H_i(t)]$. Then we have

$$H(t) = \mathbf{P}(L_i < t \leq X_i) = \mathbf{P}(L < s \leq X | L < T) = \gamma^{-1}\mathbf{P}(L < t \leq C)S(t-), \tag{4.22}$$

where $S(t-)$ is the left continuous version of $S(t)$. Hence the equation (4.21) can be written as

$$F^*(t,\boldsymbol{w}) = \int_{[a,t]}\int_{\boldsymbol{v}\leq\boldsymbol{w}} H(u) S^{-1}(u) F(du, d\boldsymbol{v}), \tag{4.23}$$

which leads to the following relationship

$$F(t,\boldsymbol{w}) = \int_{[a,t]}\int_{\boldsymbol{v}\leq\boldsymbol{w}} H^{-1}(u) S(u) F^*(du, d\boldsymbol{v}). \tag{4.24}$$

Following the idea of He and Yang (2000, 2003), we define $\boldsymbol{\Gamma} = (\zeta_{jk})_{j,k=1}^{p}$ where $\zeta_{jk} = E(W_{ij}W_{ik})$, and $\boldsymbol{q} = (\zeta_{01}, \ldots, \zeta_{0p})'$ where $\zeta_{0k} = E(X_i W_{ik})$. Multiplying both sides of the regression model in (4.2) by $\boldsymbol{W}'$ and taking expectation yields $\boldsymbol{q} = \boldsymbol{\Gamma}\boldsymbol{\beta}$. Therefore the parameter $\boldsymbol{\beta}$ can be estimated by

$$\hat{\boldsymbol{\beta}} = \hat{\boldsymbol{\Gamma}}^{-1}\hat{\boldsymbol{q}}. \tag{4.25}$$

Since $H(t)$ in (4.22) can be estimated by the empirical estimate $H_{(n)}(t) = n^{-1}\sum_{i=1}^{n} H_i(t)$ and $S(t)$ can be estimated by the product-limit estimator given in (2.10), the components of $\boldsymbol{q}$ and $\boldsymbol{\Gamma}$ can be estimated, respectively, by the unbiased estimates

$$\hat{\zeta}_{0k} = n^{-1}\sum_{i=1}^{n}\frac{X_i W_{ik}\delta_i\hat{S}(X_i-)}{H_{(n)}(X_i)}, \quad \hat{\zeta}_{jk} = n^{-1}\sum_{i=1}^{n}\frac{W_{ij}W_{ik}\delta_i\hat{S}(X_i-)}{H_{(n)}(X_i)}. \tag{4.26}$$

It can be derived that the above estimator $\hat{\boldsymbol{\beta}}$ is such a vector which minimizes $\sum_{i=1}^{n}\omega_i(X_i - \boldsymbol{W}_i\boldsymbol{\beta})^2$ where

$$\omega_i = \delta_i\hat{S}(X_i-)\Big/\sum_{j=1}^{n} I[L_j < X_i \leq X_j], \tag{4.27}$$

and hence it is the weighted least squares (WLS) estimator of $\boldsymbol{\beta}$ for LTRC data. When the event time is subject to only right censoring, it reduces to the WLS estimator in Stute (1993, 1996); He and Wong (2003). If the event time is subject to only left truncation, it reduces to the WLS estimator in He and Yang (2003).

4.3 AFT model for bivariate survival data under truncation and censoring

Consider the bivariate survival data described in Example 1.5, where the two event times are the time from hepatitis C virus (HCV) infection to referral to the clinic cohort, denoted by R, and the time from HCV infection to the development of cirrhosis, denoted by T. In clinical practice, the patients with more rapid disease progression to cirrhosis are preferentially referred to liver clinics or that referral is increasingly likely the closer a patient is to developing cirrhosis (Fu et al., 2007). The bivariate event times (R,T) are statistically correlated and the patients were included to the cohort with a so-called referral bias in epidemiology because only those who referred to the clinic before the end of study recruitment at time L were included $(R \leq L)$, i.e. R is right truncated by L. Under right truncation a patient will not be observed if the referral occurs after the end of recruitment, i.e. no information on this subject is available if $R > L$. The cirrhosis time T is subject to right censoring at time C, the last follow-up time.

Conventional univariate survival analysis, which ignores the referral bias, has been seen to produce seriously biased results in estimating the progression rate to cirrhosis within 20 to 30 years (Fu et al., 2007). To obtain an unbiased estimate of the effects of covariates on the time from HCV infection to development of cirrhosis, we consider the AFT model (4.1), where $\boldsymbol{W}$ is a random vector of covariates, β is an unknown parameter to be estimated and ε is an error term with mean 0 and an unspecified distribution function F_ε and is independent of the covariates $\boldsymbol{W}$. In the HCV data described above, the cirrhosis time T is subject to right censoring but T itself is not subject to right truncation. However, the observed sample is biased due to the truncation on the referral time R. Existing methods in survival data analysis are not readily available for the AFT model in (4.1) or (4.2) if referral bias is taken into account. Therefore in this section we will introduce a nonparametric method for the AFT model which does not assume a particular distribution of ε and does not specify a joint distribution between R and T.

4.3.1 Assumptions

Throughout this section, we make the following assumptions on $(R,T,\boldsymbol{W})$ and (L,C).

Assumption 4.3.1 *(L,C) are independent of the covariate vector $\boldsymbol{W}$.*

Assumption 4.3.2 *(L,C) are independent of the paired event times (R,T).*

Assumption 4.3.1 is reasonable for the hepatitis C study since (i) the end of recruitment was fixed before collecting the data so that L, the truncation time, can be viewed as independent of all patients' information, and (ii) the last follow-up time C can usually be any number of years after recruitment, which is also independent of patients' information. Assumption 4.3.2 is a strong assumption. Conditional independence which assumes that (L,C) and (R,T) are independent conditionally on the covariate vector $\boldsymbol{W}$ may be more appropriate in many other practical studies. Many existing methods for AFT models without considering referral bias only need this weaker assumption, see for example Miller (1976); Buckley and James (1979). Under the weaker assumption, iterative algorithms are required to locate the estimates of the regression parameter β. This is because to find the estimates we should consider the distribution of censoring variable C, which depends on the covariate vector $\boldsymbol{W}$ if conditional independence is assumed. This further implies that the distribution of residual ε depends on the unknown parameter β. Therefore, we usually allocate a starting value $\beta^{(0)}$ to calculate the initial residual distribution, and then estimate the first-step estimate $\beta^{(1)}$ based on the initial residual distribution. The final estimate of β is obtained when the convergence of the iteration algorithm is achieved. However, the stronger independence assumption that $(R,T,\boldsymbol{W})$ and (L,C) are independent allows us to work directly on the distribution of $(T,\boldsymbol{W})$ and C, and no iterative algorithm is required.

Let $G(t_1,t_2) = \mathbf{P}(L > t_1, C > t_2)$ be the continuous bivariate survival function for (L,C) and $\bar{F}(t_1,t_2) = \mathbf{P}(R \leq t_1, T \leq t_2)$ be the continuous joint distribution function for (R,T) with continuous support. We also assume the following conditions hold throughout this section.

Condition 4.3.1 *The lower boundaries of support for $\bar{F}$ are coordinate axes of the first quadrant.*

Condition 4.3.2 *For $\mathbf{t} = (t_1, t_2)$, the function $G(t_1,t_2) = \mathbf{P}(L > t_1, C > t_2) > 0$ almost surely with respect to $\bar{F}(dt_1, dt_2)$ in $\mathscr{A}$, where $\mathscr{A}$ is the common support area of $\bar{F}$ and G.*

Note that Condition 4.3.2 corresponds to Condition 3.5.3 in Chapter 3 and Condition 2.1.8 in Chapter 2 and it guarantees that $\bar{F}(t_1,t_2)$ can be identified in the area under consideration (He and Yang, 1998).

4.3.2 Estimation of regression parameters

For simplicity, we denote (R,T) as the pair of logarithm of event times, where R is the logarithm of time from HCV infection to referral to the cohort and T is the logarithm of time from infection to development of cirrhosis. The value R is randomly right-truncated at L, which is the logarithm of recruitment time. The value T is subject to random right censoring at C, which is the logarithm of the censoring time. In practice, only observations such that $R \le L$ can be collected, and the observed data are denoted by $(R_i, L_i, X_i, \delta_i, \boldsymbol{W}_i)$, $i = 1, \ldots, n$, where $X_i = \min(T_i, C_i)$ and $\delta_i = I[T_i \le C_i]$.

Using similar idea of the univariate case, let $F(t_1,t_2,\boldsymbol{w}) = \mathbf{P}(R \le t_1, T \le t_2, \boldsymbol{W} \le \boldsymbol{w})$ be the joint distribution function for $(R,T,\boldsymbol{W})$. Consider the sub-distribution function

$$\begin{aligned} F^*(t_1,t_2,\boldsymbol{w}) &= \mathbf{P}(R_i \le t_1, X_i \le t_2, \boldsymbol{W}_i \le \boldsymbol{w}_i, \delta_i = 1) \\ &= \mathbf{P}(R \le t_1, X \le t_2, \boldsymbol{W} \le \boldsymbol{w}_i, \delta = 1 | R \le L) \\ &= \gamma^{-1}\mathbf{P}(R \le t_1, T \le t_2, \boldsymbol{W} \le \boldsymbol{w}, L \ge R, C \ge T) \\ &= \gamma^{-1} \int_{[0,t_1]} \int_{[0,t_2]} \int_{\boldsymbol{u} \le \boldsymbol{w}} G(s_1-, s_2-) F(ds_1, ds_2, d\boldsymbol{u}), \end{aligned} \tag{4.28}$$

where $\gamma = \mathbf{P}(R \le L) > 0$ denotes the truncation probability. Under Condition 4.3.2, we have $G(t_1,t_2) = \mathbf{P}(L > t_1, C > t_2) > 0$. Therefore by rearranging (4.28) we have the following relation

$$F(t_1,t_2,\boldsymbol{w}) = \gamma \int_{[0,t_1]} \int_{[0,t_2]} \int_{\boldsymbol{u} \le \boldsymbol{w}} \frac{1}{G(s_1-, s_2-)} F^*(ds_1, ds_2, d\boldsymbol{u}). \tag{4.29}$$

The following lemma gives an unbiased estimating equation for $\boldsymbol{\beta}$ in (4.2), given the bivariate survival function G is known.

Lemma 4.1. *Parameter $\boldsymbol{\beta}$ in the AFT model* (4.2) *satisfies*

$$\boldsymbol{q} = \boldsymbol{\Gamma}\boldsymbol{\beta} \tag{4.30}$$

where $\boldsymbol{q} = (\zeta_{01}, \ldots, \zeta_{0p})^T$, $\zeta_{0k} = E(T_i W_{ik})$, and $\boldsymbol{\Gamma} = (\zeta_{jk})_{j,k=1}^p$, $\zeta_{jk} = E(W_{ij} W_{ik})$.

If G is known, ζ_{0k} and ζ_{jk} can be estimated, respectively, by the unbiased estimates

$$\hat{\zeta}_{0k}(G) = \frac{\hat{\gamma}}{n} \sum_{i=1}^{n} \frac{X_i W_{ik} \delta_i}{G(R_i-, X_i-)}, \quad \hat{\zeta}_{jk}(G) = \frac{\hat{\gamma}}{n} \sum_{i=1}^{n} \frac{W_{ij} W_{ik} \delta_i}{G(R_i-, X_i-)}, \tag{4.31}$$

where $\hat{\gamma}$ is the truncation probability estimate. When G is unknown, we can obtain a consistent estimate

$$\hat{\boldsymbol{\beta}} = \hat{\boldsymbol{\Gamma}}^{-1}(\hat{G})\hat{\boldsymbol{q}}(\hat{G}), \tag{4.32}$$

with $\hat{G}$ denoting a consistent estimate for the bivariate survival function $G(t_1,t_2)$. Here the truncation probability γ does not need to be estimated since it is cancelled out in (4.32).

4.3.3 Estimation of bivariate survival function G

In section 3.4, we introduced the nonparametric transformation estimator (Dai and Fu, 2012) for the bivariate survival function where both event times are subject to random right censoring and left truncation. The bivariate survival data described here can be seen as a special case where only one event time is subject to random right truncation and the other one may be subject to random right censoring. Therefore in this section, we use their method to estimate the bivariate survival function G.

Define a transformation from (t_1,t_2) to $(z;\alpha)$ as $z=\sqrt{t_1^2+t_2^2}$ and $\alpha=t_2/t_1$. For fixed α, the bivariate survival function, $G(t_1,t_2)$, can be transformed to a univariate function, $G(z;\alpha)$, by the following formula

$$G(t_1,t_2)=\mathbf{P}(L>t_1,C>t_2)=\mathbf{P}(Z(\alpha)>z):=G(z;\alpha), \tag{4.33}$$

where

$$Z(\alpha)=\min\left\{L\sqrt{1+\alpha^2},C\sqrt{1+\alpha^{-2}}\right\}. \tag{4.34}$$

In practice, the values of (L,C) may not be obtained due to truncation and censoring. Thus $Z(\alpha)$ in (4.34) may not be available. The observed data after transformation are:

$$\begin{aligned}\tilde{Z}_i(\alpha)&=\min\{\tilde{L}_i,\tilde{X}_i\},\\ \Delta_i(\alpha)&=I[\tilde{L}_i\le\tilde{X}_i]+(1-\delta_i)I[\tilde{L}_i>\tilde{X}_i],\\ V_i(\alpha)&=R_i\sqrt{1+\alpha^2},\end{aligned} \tag{4.35}$$

where $\tilde{L}_i=L_i\sqrt{1+\alpha^2}$, $\tilde{X}_i=X_i\sqrt{1+\alpha^{-2}}$ and (L_i,R_i,X_i,δ_i) are the observed original data. Such a data transformation introduces artificial censoring and truncation. For example, $\Delta_i(\alpha)=1$ implies that $\tilde{Z}_i(\alpha)$ is an observed value for $Z_i(\alpha)$, and $\Delta_i(\alpha)=0$ implies censoring. Truncation information is given by $V_i(\alpha)$.

Based on the transformed observations in (4.35), we also define

$$\begin{aligned}N(ds;\alpha)&=n^{-1}\sum_{i=1}^{n}N_i(ds;\alpha),\\ &:=n^{-1}\sum_{i=1}^{n}I\left[\tilde{Z}_i(\alpha)\in ds,s>V_i(\alpha),\Delta_i(\alpha)=1\right],\end{aligned} \tag{4.36}$$

$$\begin{aligned}H_{(n)}(s;\alpha)&=n^{-1}\sum_{i=1}^{n}H_i(s;\alpha),\\ &:=n^{-1}\sum_{i=1}^{n}I\left[\tilde{Z}_i(\alpha)>s\ge V_i(\alpha)\right],\end{aligned} \tag{4.37}$$

$$\begin{aligned}H_{(n)}(t_1,t_2)&=n^{-1}\sum_{i=1}^{n}H_i(t_1,t_2),\\ &:=n^{-1}\sum_{i=1}^{n}I[L_i>t_1\ge R_i,X_i>t_2],\end{aligned} \tag{4.38}$$

where $Z\in ds$ denotes $s\le Z<s+ds$ for simplicity. Then the following lemma leads to the estimation of the transformed univariate function $G(z;\alpha)$.

Lemma 4.2. *For fixed α, the hazard rate function of $Z(\alpha)$ is denoted by $\Lambda(dz;\alpha)=-G(dz;\alpha)/G(z-;\alpha)$. Then we have*

$$\Lambda(dz;\alpha)=\frac{\mathbf{P}\left(\tilde{Z}_i(\alpha)\in dz,z>V_i(\alpha),\Delta_i(\alpha)=1\right)}{\mathbf{P}\left(\tilde{Z}_i(\alpha)\ge z>V_i(\alpha)\right)}. \tag{4.39}$$

Proof. The lemma can be proved similarly as Lemma 3.2. ■

Lemma 4.2 implies that an estimator for $\Lambda(dz;\alpha)$ is $\hat{\Lambda}(dz;\alpha) = N(dz;\alpha)/H_{(n)}(z-;\alpha)$. Then the product-limit estimator for $G(z;\alpha)$ is

$$\hat{G}(z;\alpha) = \prod_{s\leq z}\left\{1 - \frac{N\{s;\alpha\}}{H_{(n)}(s-;\alpha)}\right\}, \tag{4.40}$$

where $N\{s;\alpha\} = N(s;\alpha) - N(s-;\alpha)$. Since $G(z;\alpha) = G(t_1,t_2)$, $\hat{G}(z;\alpha)$ in (4.40) is also an estimator for $G(t_1,t_2)$.

Define σ-field

$$\mathscr{F}^i_{z;\alpha} = \sigma\left\{I[\tilde{Z}_i(\alpha)\leq s], \Delta_i(\alpha), I[V_i(\alpha)\leq s], 0\leq s\leq z\right\}, \tag{4.41}$$

where $\tilde{Z}_i(\alpha)$, $\Delta_i(\alpha)$ and $V_i(\alpha)$ are defined in (4.35). Equipping the counting process $\{N_i(z;\alpha) : z\geq 0\}$ with the filtration $\{\mathscr{F}^i_{z;\alpha}, z\geq 0\}$, then we have

Lemma 4.3.

$$E\left[N_i(dz;\alpha)|\mathscr{F}^i_{z-;\alpha}\right] = H_i(z-;\alpha)\Lambda(dz;\alpha). \tag{4.42}$$

Proof. The lemma can be proved similarly as Lemma 3.3. ■

The following definitions, assumptions and lemma are needed to prove the large sample properties for the bivariate estimator $\hat{G}(z;\alpha)$ given in (4.40).

Definition 4.1.

1. Let $M_i(ds;\alpha) = N_i(ds;\alpha) - H_i(s-;\alpha)\Lambda(ds;\alpha)$ and $M(ds;\alpha) = n^{-1}\sum_i M_i(ds;\alpha)$. Then (4.42) implies that for fixed α, $M_i(ds;\alpha)$ and $M(ds;\alpha)$ are martingales with respect to the filtration $\mathscr{F}^i_{z;\alpha}$ and $\mathscr{F}_{z;\alpha} = \vee_{i=1}^n \mathscr{F}^i_{z;\alpha}$, respectively.
2. Let $H(t_1,t_2) := E\left[H_i(t_1,t_2)\right]$ and $H(z;\alpha) := E\left[H_i(z;\alpha)\right]$. Hence

$$\begin{aligned} H(z;\alpha) &= H(t_1,t_2) \\ &= \mathbf{P}(L_i > t_1 \geq R_i, X_i > t_2) \\ &= \mathbf{P}\big(L > t_1 \geq R, \min\{T,C\} > t_2 | L\geq R\big) \\ &= \gamma^{-1}G(t_1,t_2)\mathbf{P}(R\leq t_1, T > t_2). \end{aligned} \tag{4.43}$$

Assumption 4.3.3 *Let $\tau_\alpha = \sup_z\{z = \sqrt{t_1^2+t_2^2} : t_2/t_1 = \alpha, (t_1,t_2)\in\mathscr{A}\}$ and $\mathscr{A}$ is defined in Condition 4.3.2, we have that*

$$\sup_{\alpha\in[0,\infty]}\int_{[0,\tau_\alpha]} H(s-;\alpha)^{-1}\Lambda(ds;\alpha) < \infty. \tag{4.44}$$

Lemma 4.4. *Under Assumptions 4.3.2, 4.3.3, and Conditions 4.3.1, 4.3.2, we have that when $n\to\infty$,*

(1) $\sup_{\alpha\in[0,\infty]} E\left[\sqrt{n}\int_{[0,\tau_\alpha]} I[H_{(n)}(s-;\alpha) = 0]\Lambda(ds;\alpha)\right]^2 \to 0.$

(2) $\sup_{\alpha\in[0,\infty]} E\left[\sqrt{n}\int_{[0,\tau_\alpha]} \frac{I[H_{(n)}(s-;\alpha)=0]}{H(s-;\alpha)}\Lambda(ds;\alpha)\right]^2 \to 0.$

Proof. The lemma can be proved similarly as Lemma 2.2. ■

For $z\in[0,\tau_\alpha]$, following the results in Fleming and Harrington (1991), we have the following martingale representation for $\hat{G}(z;\alpha)$,

$$\begin{aligned} &\hat{G}(z;\alpha) - G(z;\alpha) \\ &= -G(z;\alpha)\int_{[0,z]} \frac{\hat{G}(s-;\alpha)}{G(s;\alpha)}\left[\frac{N(ds;\alpha)}{H_{(n)}(s-;\alpha)} - \Lambda(ds;\alpha)\right] \\ &= -G(z;\alpha)\int_{[0,z]} \frac{\hat{G}(s-;\alpha)}{G(s;\alpha)}\frac{I[H_{(n)}(s-;\alpha) > 0]}{H_{(n)}(s-;\alpha)}M(ds;\alpha) + B(z;\alpha), \end{aligned} \tag{4.45}$$

where

$$B(z;\alpha) = G(z;\alpha)\int_{[0,z]} \frac{\hat{G}(s-;\alpha)}{G(s;\alpha)} I[H_{(n)}(s-;\alpha) > 0]\Lambda(ds;\alpha). \tag{4.46}$$

Hence we have,

$$\begin{aligned}
&\sqrt{n}[\hat{G}(z;\alpha) - G(z;\alpha)] \\
&= -\sqrt{n}G(z;\alpha)\int_{[0,z]} \Big[\frac{\hat{G}(s-;\alpha)}{G(s;\alpha)}\frac{I[H_{(n)}(s-;\alpha) > 0]}{H_{(n)}(s-;\alpha)} - \frac{1}{H(s-;\alpha)}\Big] M(ds;\alpha) \\
&\quad - \sqrt{n}G(z;\alpha)\int_{[0,z]} \frac{1}{H(s-;\alpha)} M(ds;\alpha) + \sqrt{n}B(z;\alpha) \\
&:= \omega_n(z;\alpha) - \sqrt{n}G(z;\alpha)\int_{[0,z]} \frac{1}{H(s-;\alpha)} M(ds;\alpha) + \sqrt{n}B(z;\alpha).
\end{aligned} \tag{4.47}$$

To prove the consistency of $\hat{G}(z;\alpha)$, we need to prove that each term in (4.47) converges to 0 as $n \to \infty$.

Lemma 4.5. *The first term in* (4.47) *satisfies*

$$\sup_{\alpha\in[0,\infty], z\in[0,\tau_\alpha]} E\,[\omega_n(z;\alpha)]^2 \to 0. \tag{4.48}$$

■

Proof. The proof is similar to Lemma 4.4.

Since $G(z;\alpha) \le G(s;\alpha)$ given $z \ge s$, we have $B(z;\alpha) \le \int_{[0,z]} I[H_{(n)}(s-;\alpha) = 0]\Lambda(ds;\alpha)$ almost surely. Hence using the results of Lemma 4.4 we have that when $n \to \infty$,

$$\sup_{\alpha\in[0,\infty], z\in[0,\tau_\alpha]} E\left[\sqrt{n}B(z;\alpha)\right]^2 \to 0. \tag{4.49}$$

In addition,we have that

$$E\left[\int_{[0,z]} \frac{1}{H(s-;\alpha)} M(ds;\alpha)\right]^2 = n^{-1}\int_{[0,z]} \frac{1}{H(s-;\alpha)} \Lambda(ds;\alpha) = O(n^{-1}), \tag{4.50}$$

where the last equality sign can be obtained from Assumption 4.3.3.

Using the results in Lemma 4.5, together with (4.47), (4.49) and (4.50), we have that when $n \to \infty$,

$$E\left[\hat{G}(z;\alpha) - G(z;\alpha)\right]^2 \to 0, \tag{4.51}$$

which indicates the consistency of the estimator $\hat{G}(z;\alpha)$.

Theorem 4.1. *Under Assumptions 4.3.2, 4.3.3, and Conditions 4.3.1, 4.3.2, for any* $(t_1,t_2) \in \mathscr{A}$, *we have*

$$\sqrt{n}[\hat{G}(z;\alpha) - G(z;\alpha)] \Rightarrow N(0, \sigma_G^2(z;\alpha)), \tag{4.52}$$

where the asymptotic variance

$$\sigma_G^2(z;\alpha) = G^2(z;\alpha)\int_{[0,z]} \frac{1}{H(s-;\alpha)} \Lambda(ds;\alpha). \tag{4.53}$$

Proof. From (4.47) and Definition (4.1), we have that

$$\sqrt{n}[\hat{G}(z;\alpha) - G(z;\alpha)] = -n^{1/2}G(z;\alpha)\sum_{j=1}^{n}\int_{[0,z]} \frac{1}{H(s-;\alpha)} M_j(ds;\alpha) + r_n(z;\alpha), \tag{4.54}$$

where $r_n(z;\alpha) = \omega_n(z;\alpha) + \sqrt{n}B(z;\alpha)$ and $\sup_{\alpha\in[0,\infty],z\in[0,\tau_\alpha]} E[r_n(z;\alpha)]^2 \to 0$. Then the theorem follows using the idea in Dai and Fu (2012). ∎

Theorem 4.1 implies the asymptotic normality of $\hat{G}(z;\alpha)$. Substituting the consistent estimators $\hat{\Lambda}(z;\alpha)$ and $\hat{G}(z;\alpha)$ into (4.53), we can obtain a consistent estimator for $\sigma_G^2(z;\alpha)$ which is

$$\hat{\sigma}_G^2(z;\alpha) = \hat{G}^2(z;\alpha)\int_{[0,z]} \frac{I[H_{(n)}(s-;\alpha) > 0]}{H_{(n)}(s-;\alpha)}\hat{\Lambda}(ds;\alpha). \tag{4.55}$$

4.3.4 Large sample properties of $\hat{\beta}$

The consistency of the regression parameter estimate $\hat{\beta}$ in (4.32) is implied in the following theorem.

Theorem 4.2. *Let β^* denote the true value of the regression parameter β in the AFT model* (4.2)*. Then for the estimator $\hat{\beta}$ given in* (4.32)*, we have*

$$\lim_{n\to\infty} \hat{\beta} = \beta^*$$

in probability, i.e. $\hat{\beta}$ is consistent.

Proof. The proof is a simple argument based on the Lemma 8.3.1 in Fleming and Harrington (1991). Let a sequence of concave functions be of the form

$$\begin{aligned}
F_n(\beta) &= n^{-1}\sum_{i=1}^{n} \frac{\delta_i I[\hat{G}(R_i-,X_i-) > 0]}{\hat{G}(R_i-,X_i-)}(X_i - \mathbf{W}_i\beta)^2 \\
&= n^{-1}\sum_{i=1}^{n} \frac{\delta_i I[\hat{G}(R_i-,X_i-) > 0]}{\hat{G}(R_i-,X_i-)}(X_i - \mathbf{W}_i\beta^*)^2 \\
&\quad + 2n^{-1}\sum_{i=1}^{n} \frac{\delta_i I[\hat{G}(R_i-,X_i-) > 0]}{\hat{G}(R_i-,X_i-)}(X_i - \mathbf{W}_i\beta^*)(\mathbf{W}_i\beta^* - \mathbf{W}_i\beta) \\
&\quad + n^{-1}\sum_{i=1}^{n} \frac{\delta_i I[\hat{G}(R_i-,X_i-) > 0]}{\hat{G}(R_i-,X_i-)}(\mathbf{W}_i\beta^* - \mathbf{W}_i\beta)^2 \\
&:= I + II + III,
\end{aligned} \tag{4.56}$$

where the term II can also be written as

$$\begin{aligned}
II &= 2n^{-1}\sum_{i=1}^{n}\left[\frac{\delta_i I[\hat{G}(R_i-,X_i-) > 0]}{\hat{G}(R_i-,X_i-)} - \frac{\delta_i}{G(R_i-,X_i-)}\right](X_i - \mathbf{W}_i\beta^*)(\mathbf{W}_i\beta^* - \mathbf{W}_i\beta) \\
&\quad + 2n^{-1}\sum_{i=1}^{n} \frac{\delta_i}{G(R_i-,X_i-)}(X_i - \mathbf{W}_i\beta^*)(\mathbf{W}_i\beta^* - \mathbf{W}_i\beta) \\
&:= II_{(1)} + II_{(2)}.
\end{aligned} \tag{4.57}$$

The term $II_{(1)} \to 0$ as $n \to \infty$ since

$$\frac{I[\hat{G}(R_i-,X_i-) > 0]}{\hat{G}(R_i-,X_i-)} - \frac{1}{G(R_i-,X_i-)} \xrightarrow{P} 0. \tag{4.58}$$

Meanwhile, for the term $II_{(2)}$, we have as $n \to \infty$

$$
\begin{aligned}
II_{(2)} &\to E\left[\frac{\delta_i}{G(R_i-,X_i-)}(X_i - \boldsymbol{W}_i\boldsymbol{\beta}^*)(\boldsymbol{W}_i\boldsymbol{\beta}^* - \boldsymbol{W}_i\boldsymbol{\beta})\right] \\
&= E_{\boldsymbol{W}_i}\left\{E\left[\frac{\delta_i}{G(R_i-,X_i-)}(X_i - \boldsymbol{W}_i\boldsymbol{\beta}^*)|\boldsymbol{W}_i\right]\cdot(\boldsymbol{W}_i\boldsymbol{\beta}^* - \boldsymbol{W}_i\boldsymbol{\beta})\right\} \\
&= E_{\boldsymbol{W}_i}\left\{E\left[\varepsilon\right]\cdot(\boldsymbol{W}_i\boldsymbol{\beta}^* - \boldsymbol{W}_i\boldsymbol{\beta})\right\} \\
&= 0,
\end{aligned} \tag{4.59}
$$

since we assume that the error term ε_i has mean 0. Hence the term II in (4.56) converges to 0 as $n \to \infty$.

Using the similar idea above, we can also have that as $n \to \infty$, the term $I \to E[\varepsilon^2] = Var[\varepsilon]$, which is a constant. In addition, the term III is a concave function which reaches the minimum at $\boldsymbol{\beta} = \boldsymbol{\beta}^*$. Let $\lim_{n\to\infty} F_n(\boldsymbol{\beta}) = f(\boldsymbol{\beta})$. Then the function $f(\boldsymbol{\beta})$ has a unique minimum value at $\boldsymbol{\beta} = \boldsymbol{\beta}^*$.

Using the results of Lemma 8.3.1 in Fleming and Harrington (1991), together with the case that $F_n(\boldsymbol{\beta})$ reaches its unique minimum at $\boldsymbol{\beta} = \hat{\boldsymbol{\beta}}$, we can conclude that $\hat{\boldsymbol{\beta}}$ converges to the true value $\boldsymbol{\beta}^*$ in probability as $n \to \infty$. ∎

The estimator $\hat{\boldsymbol{\beta}}$ obtained from (4.32) is also equivalent to solving the estimating equation

$$
\boldsymbol{Q}(\boldsymbol{\beta};\hat{G}) = n^{-1}\sum_{i=1}^{n}\frac{\delta_i I[\hat{G}(R_i-,X_i-) > 0]}{\hat{G}(R_i-,X_i-)}(X_i - \boldsymbol{W}_i\boldsymbol{\beta})\boldsymbol{W}_i^T = \boldsymbol{0}, \tag{4.60}
$$

since $\boldsymbol{Q}(\hat{\boldsymbol{\beta}};\hat{G}) = \boldsymbol{\Gamma}(\hat{G})\hat{\boldsymbol{\beta}} - \boldsymbol{q}(\hat{G}) = \boldsymbol{0}$. Then the following theorem provides the results of the asymptotic normality of $\hat{\boldsymbol{\beta}}$.

Theorem 4.3. *Let*

$$
\boldsymbol{\eta}_i = \delta_i(X_i - \boldsymbol{W}_i\boldsymbol{\beta}^*)\boldsymbol{W}_i^T, \quad \xi_{ji} = \int_{[0,\check{Z}_i(\alpha_i)]}\frac{1}{H(s-;\alpha_i)}M_j(ds;\alpha_i), \tag{4.61}
$$

where $\alpha_i = X_i/R_i$. Let $\boldsymbol{Q}'(\boldsymbol{\beta};G) = \partial\boldsymbol{Q}(\boldsymbol{\beta};G)/\partial\boldsymbol{\beta}$ and $\mathscr{D}_k = \{R_k, L_k, X_k, \delta_k, \boldsymbol{W}_k\}$ denotes the observed information for the subject k. Also denote $\boldsymbol{\beta}^*$ as the true value of the regression parameter vector $\boldsymbol{\beta}$. Then we have that

$$
\sqrt{n}(\hat{\boldsymbol{\beta}} - \boldsymbol{\beta}^*) \Rightarrow N(\boldsymbol{0}, \boldsymbol{\Sigma}_{\boldsymbol{\beta}}), \tag{4.62}
$$

where $\boldsymbol{\Sigma}_{\boldsymbol{\beta}} = \left[\boldsymbol{Q}'(\boldsymbol{\beta};G)\right]^{-1}\boldsymbol{\Sigma}_{\boldsymbol{Q}}\left\{\left[\boldsymbol{Q}'(\boldsymbol{\beta};G)\right]^{-1}\right\}^T$. The matrix $\boldsymbol{\Sigma}_{\boldsymbol{Q}}$ is given by

$$
\boldsymbol{\Sigma}_{\boldsymbol{Q}} = Var\left[\frac{\boldsymbol{\eta}_k}{G(R_k-,X_k-)} + \boldsymbol{\Phi}(\mathscr{D}_k)\right], \tag{4.63}
$$

$$
\boldsymbol{\Phi}(\mathscr{D}_k) = E\left\{G^{-1}(R_i-,X_i-)\xi_{ki}\boldsymbol{\eta}_i|\mathscr{D}_k\right\}, \; i \neq k. \tag{4.64}
$$

∎

Proof. We start from representing $\sqrt{n}\boldsymbol{Q}(\boldsymbol{\beta}^*;\hat{G})$ as

$$
\sqrt{n}\boldsymbol{Q}(\boldsymbol{\beta}^*;\hat{G}) = \sqrt{n}\boldsymbol{Q}(\boldsymbol{\beta}^*;G) + \sqrt{n}\left[\boldsymbol{Q}(\boldsymbol{\beta}^*;\hat{G}) - \boldsymbol{Q}(\boldsymbol{\beta}^*;G)\right]. \tag{4.65}
$$

The first term of the right side of (4.65) is a sum of i.i.d random variables since

$$
\sqrt{n}\boldsymbol{Q}(\boldsymbol{\beta}^*;G) = n^{-1/2}\sum_{i=1}^{n}\frac{\delta_i}{G(R_i-,X_i-)}(X_i - \boldsymbol{W}_i\boldsymbol{\beta}^*)\boldsymbol{W}_i^T.
$$

The second term of the right side of (4.65) can be written as

$$
\begin{aligned}
&\sqrt{n}\big[Q(\beta^*;\hat{G})-Q(\beta^*;G)\big]\\
&=n^{-1/2}\sum_{i=1}^{n}\left[\frac{I[\hat{G}(R_i-,X_i-)>0]}{\hat{G}(R_i-,X_i-)}-\frac{1}{G(R_i-,X_i-)}\right]\delta_i(X_i-\boldsymbol{W}_i\beta^*)\boldsymbol{W}_i^T\\
&=n^{-1/2}\sum_{i=1}^{n}\frac{G(R_i-,X_i-)-\hat{G}(R_i-,X_i-)}{G^2(R_i-,X_i-)}\eta_i\\
&\quad+n^{-1/2}\sum_{i=1}^{n}\frac{G(R_i-,X_i-)-\hat{G}(R_i-,X_i-)}{G(R_i-,X_i-)}\left[\frac{I[\hat{G}(R_i-,X_i-)>0]}{\hat{G}(R_i-,X_i-)}-\frac{1}{G(R_i-,X_i-)}\right]\eta_i\\
&:=I+II.
\end{aligned}\tag{4.66}
$$

Using the results in (4.54), we have that as $n\to\infty$, II in (4.66) converges to 0 in probability. Hence, based on the observed transformed data defined in (4.35) and the results in (4.54), we have that

$$
\begin{aligned}
\sqrt{n}\big[Q(\beta^*;\hat{G})-Q(\beta^*;G)\big]&=n^{-3/2}\sum_{i=1}^{n}\frac{\eta_i}{G(\tilde{Z}_i(\alpha_i);\alpha_i)}\left[\sum_{j=1}^{n}\int_{[0,\tilde{Z}_i(\alpha_i)]}\frac{1}{H(s-;\alpha_i)}M_j(ds;\alpha_i)\right]+o_p(1)\\
&=n^{-3/2}\sum_{i=1}^{n}\sum_{j=1}^{n}\frac{\xi_{ji}\eta_i}{G(\tilde{Z}_i(\alpha_i);\alpha_i)}+o_p(1):=\boldsymbol{U}_n+o_p(1).
\end{aligned}\tag{4.67}
$$

According to the properties of U-statistics (Serfling, 1980), we have $\boldsymbol{U}_n=\boldsymbol{U}_n+o\big(n^{-1}(\log n)^{\rho}\big)$ for some $\rho>0$, where

$$
\begin{aligned}
\boldsymbol{U}_n&=\sum_{k=1}^{n}E\big[\boldsymbol{U}_n\,|\,\mathscr{D}_k\big]=\sum_{k=1}^{n}E\Big[n^{-3/2}\sum_{i,j=1}^{n}G^{-1}(R_i-,X_i-)\xi_{ji}\eta_i\,|\,\mathscr{D}_k\Big]\\
&=n^{-3/2}\sum_{i,j,k=1}^{n}E\big[G^{-1}(R_i-,X_i-)\xi_{ji}\eta_i\big|\mathscr{D}_k\big].
\end{aligned}\tag{4.68}
$$

Since ξ_{ji}, defined in (4.61), is a zero-martingale given $\mathscr{D}_i$, we then have $E\{\xi_{ji}|\mathscr{D}_i\}=0$. Hence if $j\neq k$, $E\big[G^{-1}(R_i-,X_i-)\xi_{ji}\eta_i\big|\mathscr{D}_k\big]=E\big[G^{-1}(R_i-,X_i-)\xi_{ji}\eta_i\big]=\mathbf{0}$. Hence we have

$$
\begin{aligned}
\hat{\boldsymbol{U}}_n&=n^{-1/2}\sum_{k=1}^{n}\frac{n-1}{n}E\left\{G^{-1}(R_i-,X_i-)\xi_{ki}\eta_i\,|\,\mathscr{D}_k\right\}\\
&=n^{-1/2}\sum_{k=1}^{n}E\left\{G^{-1}(R_i-,X_i-)\xi_{ki}\eta_i\,|\,\mathscr{D}_k\right\}+o_p(1)\\
&=n^{-1/2}\sum_{k=1}^{n}\boldsymbol{\Phi}(\mathscr{D}_k)+o_p(1).
\end{aligned}\tag{4.69}
$$

The above equation, together with (4.67), implies that $\sqrt{n}\big[Q(\beta^*;\hat{G})-Q(\beta^*;G)\big]=n^{-1/2}\sum_{k=1}^{n}\boldsymbol{\Phi}(\mathscr{D}_k)+o_p(1)$, which is a sum of i.i.d. terms. Then the variance-covariance matrix of $\sqrt{n}Q(\beta^*;\hat{G})$ is given by

$$
\begin{aligned}
\boldsymbol{\Sigma}_Q=\mathrm{Var}\big[\sqrt{n}Q(\beta^*;\hat{G})\big]&=\mathrm{Var}\Big[\sqrt{n}Q(\beta^*;G)+\sqrt{n}\big[Q(\beta^*;\hat{G})-Q(\beta^*;G)\big]\Big]\\
&=\mathrm{Var}\left[\frac{\eta_k}{G(R_k-,X_k-)}+\boldsymbol{\Phi}(\mathscr{D}_k)\right],\quad k=1,\ldots,n.
\end{aligned}
$$

Then the theorem follows from the first-order Taylor expansion for a vector field

$$
Q(\beta^*;\hat{G})\approx Q(\hat{\beta};\hat{G})+Q'(\hat{\beta};\hat{G})(\beta^*-\hat{\beta})\tag{4.70}
$$

■

4.3.5 Simulation studies

In this section, we conduct simulation studies to assess the performance of the proposed regression parameter estimator given in (4.32). Truncation times L and censoring times C are generated respectively from

$$C = a\nu_1 + b\nu_2,$$
$$L = c\nu_1 + d\nu_2 + U[0,1],$$

where ν_1 and ν_2 are exponentially distributed with unit mean. Difference censoring percentages and truncation probabilities can be achieved by adjusting the values of (a,b,c,d). Pairs of event times are generated from the following model to mimic the real data in the Example 1.5. The logarithm of event time T follows $T = \boldsymbol{W}\beta^* + \varepsilon$, where the true value $\beta^* = (3.7, -0.05, -0.3, -0.1)^T$ and the covariate $\boldsymbol{W}$ is defined as: the intercept $W_1 = 1$ and the predictors $W_2 \sim U[20,30]$, $W_3 \sim$ Bernoulli(0.5) and $W_4 \sim$ Bernoulli(0.5). The observed sample size is set at $n = 200$ and 100. The replication time is 500. The time R has mean $ER = 1.35$ and (R,T) are correlated through a joint distribution of $(R - ER, \varepsilon)$. Once we simulate $(R - ER, \varepsilon)$ we then have the simulated values for R and $T = \boldsymbol{W}\beta^* + \varepsilon$. We here consider two types of joint distribution functions of $(R - ER, \varepsilon)$.

Scenario 1: The two error terms $R - ER$ and ε are generated as follows,

$$R - ER = 1.0\upsilon_1 + 0.5\upsilon_2,$$
$$\varepsilon = 0.4\upsilon_1 + 0.35\upsilon_2,$$

where υ_1 and υ_2 are two uniform random variables from $\mathbf{U}[-0.5, 0.5]$. The linear combinations above can make $R - ER$ and ε correlated. The simulation results are presented in Table 4.1. We choose different values for (a,b,c,d) to achieve different censoring and truncation percentages.

When sample size is large ($n = 200$), censoring percentage is low (about 20%) and truncation probability is high (about 0.85), the simulation results indicate that the biases are 0.001 for all parameters, which are very small. In this case, the mean standard deviation estimate and standard deviation for Monte Carlo estimates are very close. Even when the data are severely biased (80% censoring and 0.15 truncation probability), the biases for the predictor parameters $(\beta_2, \beta_3, \beta_4)$ are still very small, (0.001, 0.010, 0.004) respectively, although the bias for intercept β_1 is larger. Therefore we can conclude that the proposed estimators work well for large sample sizes. Note that when sample size is $n = 100$, censoring percentage is about 80% and truncation probability is about 0.15, the estimate for the intercept has the largest biases 0.064 and the estimates for other parameters are still good. In this case the standard error based on Monte Carlo simulations $\hat{s}_\beta$ and the mean of estimated standard errors $\hat{\sigma}_\beta$ are not close. This is not surprising as the observed sample is severely biased (truncation probability is only about 0.15) and among the observed samples 80% are censored, i.e. only about 20 observations are fully observed.

Scenario 2: Instead of generating the error terms from linear combinations, we generate $R - ER$ and ε as follows. First generate (ω_1, ω_2) from the well-known bivariate parametric model in Clayton (1978), which has joint survival function $S_\varepsilon(s_1, s_2) = (S_{\varepsilon_1}(s_1)^{-\phi} + S_{\varepsilon_2}(s_2)^{-\phi} - 1)^{-1/\phi}$. We take $\phi = 4$ (see for example Prentice et al., 2004). The marginal survival functions $S_{\varepsilon_i}(s_i)$ are from unit exponential distribution truncated at 1 and its mean is $E = (1 - 2e^{-1})/(1 - e^{-1})$. Then we let $R = ER + \omega_1 - E$ and $\varepsilon = \omega_2 - E$. We can also choose different values for (a,b,c,d) to achieve different censoring and truncation percentages. The simulation results are summarized in Table 4.2. We have similar findings and conclusions as for Scenario 1.

Table 4.1: Scenario 1. (e) estimate (bias in parenthesis); β^*: true value of β; n: observed sample size; γ: truncation probability; $c\%$: censoring percentage; $\hat{s}_\beta$: means of standard deviation estimates obtained by Theorem 4.3; $\hat{\sigma}_\beta$: the standard deviation for $\hat{\beta}$ and $\hat{\beta}$ based on the 500 simulations.

		c% = 20%			c% = 50%			c% = 80%		
	β^*	(e)	$\hat{s}_\beta$	$\hat{\sigma}_\beta$	(e)	$\hat{s}_\beta$	$\hat{\sigma}_\beta$	(e)	$\hat{s}_\beta$	$\hat{\sigma}_\beta$
$n=200$										
$\gamma=0.85$	3.700	3.699(0.001)	0.115	0.112	3.717(0.017)	0.160	0.156	3.749(0.049)	0.269	0.246
	-0.050	-0.051(0.001)	0.004	0.004	-0.051(0.001)	0.006	0.006	-0.051(0.001)	0.010	0.009
	-0.300	-0.301(0.001)	0.024	0.025	-0.305(0.005)	0.035	0.035	-0.304(0.004)	0.068	0.059
	-0.100	-0.099(0.001)	0.025	0.025	-0.103(0.003)	0.035	0.034	-0.102(0.002)	0.058	0.055
$\gamma=0.5$	3.700	3.708(0.008)	0.126	0.131	3.748(0.048)	0.176	0.171	3.741(0.041)	0.317	0.272
	-0.050	-0.050(0.000)	0.005	0.005	-0.051(0.001)	0.007	0.007	-0.051(0.001)	0.012	0.010
	-0.300	-0.305(0.005)	0.029	0.028	-0.304(0.004)	0.038	0.037	-0.308(0.008)	0.068	0.058
	-0.100	-0.100(0.000)	0.029	0.028	-0.101(0.001)	0.036	0.035	-0.103(0.003)	0.058	0.055
$\gamma=0.15$	3.700	3.649(0.051)	0.217	0.183	3.755(0.055)	0.255	0.198	3.759(0.059)	0.363	0.278
	-0.050	-0.049(0.001)	0.008	0.007	-0.050(0.001)	0.010	0.008	-0.051(0.001)	0.014	0.010
	-0.300	-0.287(0.013)	0.054	0.043	-0.298(0.002)	0.056	0.044	-0.290(0.010)	0.080	0.058
	-0.100	-0.099(0.001)	0.058	0.043	-0.102(0.002)	0.056	0.043	-0.096(0.004)	0.076	0.055
$n=100$										
$\gamma=0.85$	3.700	3.725(0.025)	0.162	0.161	3.726(0.026)	0.241	0.216	3.665(0.335)	0.446	0.325
	-0.050	-0.051(0.001)	0.006	0.006	-0.050(0.000)	0.009	0.008	-0.048(0.002)	0.017	0.012
	-0.300	-0.301(0.001)	0.039	0.037	-0.302(0.002)	0.055	0.051	-0.295(0.005)	0.102	0.073
	-0.100	-0.100(0.000)	0.037	0.035	-0.100(0.000)	0.049	0.048	-0.107(0.007)	0.094	0.071
$\gamma=0.5$	3.700	3.710(0.010)	0.180	0.180	3.664(0.036)	0.250	0.229	3.726(0.026)	0.429	0.338
	-0.050	-0.051(0.001)	0.007	0.007	-0.052(0.002)	0.009	0.009	-0.051(0.001)	0.015	0.012
	-0.300	-0.304(0.004)	0.041	0.040	-0.308(0.008)	0.053	0.050	-0.292(0.008)	0.099	0.069
	-0.100	-0.098(0.002)	0.040	0.040	-0.102(0.002)	0.053	0.049	-0.101(0.001)	0.091	0.069
$\gamma=0.15$	3.700	3.638(0.062)	0.266	0.202	3.655(0.045)	0.320	0.222	3.764(0.064)	0.528	0.343
	-0.050	-0.049(0.001)	0.010	0.008	-0.048(0.002)	0.012	0.009	-0.049(0.001)	0.020	0.013
	-0.300	-0.293(0.007)	0.063	0.048	-0.295(0.005)	0.067	0.050	-0.285(0.015)	0.109	0.067
	-0.100	-0.099(0.001)	0.067	0.047	-0.091(0.009)	0.069	0.051	-0.094(0.006)	0.097	0.062

4.3.6 Example of data analysis

We illustrate the proposed method with the Edinburgh hepatitis C data in Fu et al. (2007). The patients were studied retrospectively and followed prospectively for the development of HCV-related cirrhosis. HCV patients usually experience no symptoms or mild symptoms in the early stages and are often referred to hospital shortly before they develop cirrhosis or complications. Among these individuals, there is no cirrhosis event occurred prior to their referral time and 63 (16%) developed cirrhosis during the follow-up. The median duration time from HCV infection to referral is 17.1 years and the median follow-up time from referral to cirrhosis or censoring is 2.4 years.

Our aim is to determine how the progression to cirrhosis is affected by the three covariates: age at infection, HIV co-infection (yes:1 or no:0) and heavy alcohol consumption (yes:1 or no:0). An individual with heavy alcohol intake was defined as one consuming more than 50 units alcohol per week for at least 5 years.

Table 4.2: Scenario 2. (e): estimate (bias in parenthesis); β^*: true value of β; n: observed sample size; γ: truncation probability; $c\%$: censoring percentage; $\hat{s}_\beta$: means of standard deviation estimates obtained by Theorem 4.3; $\hat{\sigma}_\beta$: the standard deviation for $\hat{\beta}$ and $\hat{\beta}$ based on the 500 simulations.

		c% = 20%			c% = 50%			c% = 80%		
	β^*	(e)	$\hat{s}_\beta$	$\hat{\sigma}_\beta$	(e)	$\hat{s}_\beta$	$\hat{\sigma}_\beta$	(e)	$\hat{s}_\beta$	$\hat{\sigma}_\beta$
$n = 200$										
$\gamma = 0.85$	3.700	3.665(0.005)	0.220	0.208	3.621(0.079)	0.307	0.286	3.573(0.137)	0.427	0.377
	-0.050	-0.049(0.001)	0.008	0.008	-0.049(0.001)	0.012	0.011	-0.048(0.002)	0.016	0.015
	-0.300	-0.298(0.002)	0.048	0.047	-0.289(0.011)	0.062	0.061	-0.279(0.021)	0.096	0.087
	-0.100	-0.099(0.001)	0.047	0.046	-0.102(0.002)	0.063	0.061	-0.093(0.007)	0.095	0.086
$\gamma = 0.5$	3.700	3.646(0.054)	0.218	0.212	3.637(0.063)	0.262	0.248	3.561(0.139)	0.430	0.382
	-0.050	-0.050(0.000)	0.008	0.008	-0.049(0.001)	0.010	0.010	-0.048(0.002)	0.016	0.015
	-0.300	-0.290(0.010)	0.047	0.048	-0.296(0.004)	0.060	0.060	-0.274(0.026)	0.010	0.090
	-0.100	-0.096(0.004)	0.049	0.048	-0.105(0.005)	0.060	0.058	-0.095(0.005)	0.095	0.088
$\gamma = 0.15$	3.700	3.590(0.110)	0.319	0.296	3.547(0.153)	0.346	0.316	3.514(0.186)	0.441	0.391
	-0.050	-0.048(0.002)	0.013	0.012	-0.047(0.003)	0.014	0.013	-0.046(0.004)	0.017	0.015
	-0.300	-0.289(0.011)	0.070	0.068	-0.278(0.022)	0.080	0.076	-0.265(0.035)	0.111	0.089
	-0.100	-0.093(0.007)	0.073	0.068	-0.094(0.006)	0.082	0.074	-0.098(0.002)	0.105	0.089
$n = 100$										
$\gamma = 0.85$	3.700	3.640(0.060)	0.296	0.295	3.603(0.097)	0.404	0.353	3.480(0.220)	0.538	0.472
	-0.050	-0.049(0.001)	0.011	0.011	-0.048(0.002)	0.016	0.014	-0.045(0.005)	0.022	0.019
	-0.300	-0.288(0.012)	0.066	0.066	-0.293(0.007)	0.089	0.085	-0.281(0.019)	0.140	0.112
	-0.100	-0.098(0.002)	0.064	0.066	-0.095(0.005)	0.089	0.085	-0.081(0.019)	0.133	0.112
$\gamma = 0.5$	3.700	3.627(0.073)	0.376	0.355	3.613(0.087)	0.397	0.359	3.496(0.204)	0.635	0.500
	-0.050	-0.049(0.001)	0.014	0.014	-0.048(0.002)	0.016	0.014	-0.046(0.004)	0.024	0.019
	-0.300	-0.283(0.017)	0.072	0.071	-0.293(0.007)	0.084	0.083	-0.271(0.029)	0.137	0.113
	-0.100	-0.098(0.001)	0.072	0.067	-0.099(0.001)	0.086	0.084	-0.082(0.018)	0.139	0.111
$\gamma = 0.15$	3.700	3.512(0.188)	0.384	0.326	3.494(0.206)	0.480	0.412	3.434(0.266)	0.647	0.537
	-0.050	-0.046(0.004)	0.015	0.013	-0.045(0.005)	0.018	0.016	-0.044(0.006)	0.025	0.021
	-0.300	-0.265(0.035)	0.089	0.079	-0.267(0.033)	0.109	0.089	-0.260(0.040)	0.141	0.108
	-0.100	-0.095(0.005)	0.093	0.081	-0.093(0.007)	0.105	0.090	-0.078(0.022)	0.136	0.110

Table 4.3: Estimation results (SE in parenthesis), * means the estimate is significant at 5% level.

	With truncation – $\hat{\beta}$ (SE)	Without truncation – $\hat{\beta}$ (SE)
Intercept	3.628* (0.068)	3.844* (0.133)
Age at infection	-0.011* (0.003)	-0.031* (0.004)
HIV co-infection	-0.313* (0.047)	-0.380* (0.089)
Heavy alcohol consumption	-0.098* (0.038)	-0.077 (0.070)

Table 4.3 summarizes the estimates of regression parameters obtained from our method. The results from the truncated model, where the referral bias is considered, show that age at infection, HIV co-infection and heavy alcohol in-take are significantly identified as risk factors associated with more rapid disease progression. If we compare the results with those from a non-truncated model, ignorance of the referral bias has failed to identify heavy alcohol consumption as a significant risk factor. In medical literatures, older age at infection, HIV co-infection and heavy alcohol intake have all been identified as factors associated with more rapid hepatitis C disease progression (Sharma and Sherker, 2009).

In Table 4.3, although the proposed methods do make a difference in terms of significance of the covariate 'Heavy alcohol consumption', the resulting estimated values and the standard errors by the two models seem to be very similar. This is because in the Edinburgh hepatitis C data in Fu et al. (2007), 387 patients are observed. Among these individuals, only 63 (16%) developed cirrhosis during follow-up. Therefore the censoring percentage is around 85%, which is quite high. The inverse probability weighted (IPW) estimator of the bivariate survival function G used in our WLS method does struggle when the censoring percentage is high (Dai and Bao, 2009; Dai and Fu, 2012). Therefore the insignificance of the improvement when analysing the severely biased survival data (with high censoring percentage) is reasonable.

Based on our coefficient estimates shown in Table 4.3, we use $\hat{T} = \exp(\boldsymbol{W}\boldsymbol{\beta})$ to predict the time period from infection to cirrhosis for individuals with different values of covariates. The range of age at infection is taken to be from 10 to 70, since in the Edinburgh hepatitis C data only 5 of 387 patients were infected by HCV before 10 years old. The prediction results according to different values of covariates are shown in Figure 4.1. We can see that with the referral bias taken into account, when the age at infection is greater than 10, the predicted durations from infection to cirrhosis are longer compared to the case without considering referral bias. This is because that the patients with more rapid disease progression are preferentially referred to the liver clinic cohort. Hence the time period from infection to cirrhosis observed in the clinic cohort may be shorter than that for the whole HCV patients community (Wang et al., 2013). If so, removing the referral bias, the predicted values should be larger comparing with the case only censoring is involved, as revealed by Fu et al. (2007).

4.3.7 Model diagnostic

To assess the linearity assumption of the AFT model, we consider using the χ^2 goodness-of-fit test. Following the idea in Kim (1993), we use a revised χ^2 statistic for the goodness-of-fit test for models with incomplete data. To calculate the test statistic, support of the response variable (logarithm of the observed time to cirrhosis) is partitioned into several subintervals and then the observed and expected numbers of observations in each subinterval are worked out.

Specifically, we partition the logarithm of the observed time to cirrhosis into $k+1$ subintervals. Let F^{KM} be the KM estimator based on the observed cirrhosis time X and the censoring indicator, and let $\tilde{F}^{\text{KM}}$ be the KM estimator based on the error term $\varepsilon = X - \boldsymbol{W}\hat{\boldsymbol{\beta}}$. Under the null hypothesis we calculate the χ^2 test statistic

$$\sum_{i=1}^{k+1} \frac{(O_i - E_i)^2}{\text{variance}}, \quad i = 1, \ldots, k+1, \tag{4.71}$$

where $O_i = n(F_i^{\text{KM}} - F_{i-1}^{\text{KM}})$ is the observed number of observations in the ith subinterval, $E_i = n(\tilde{F}_i^{\text{KM}} - \tilde{F}_{i-1}^{\text{KM}})$ is the expected number of observations in the ith subinterval, and n is the sample size of our cirrhosis data ($n = 387$). The variance is calculated by

$$\text{variance} = \sum_{i=1}^{k+1} \int \frac{dF_i^{\text{KM}}}{(1 - H_i)(1 - F_i^{\text{KM}})}, \tag{4.72}$$

where $H(x)$ is the distribution function of $X = \min(T, C)$, and $1 - H(x)$ can be estimated by the proportion that $X \geqslant x$.

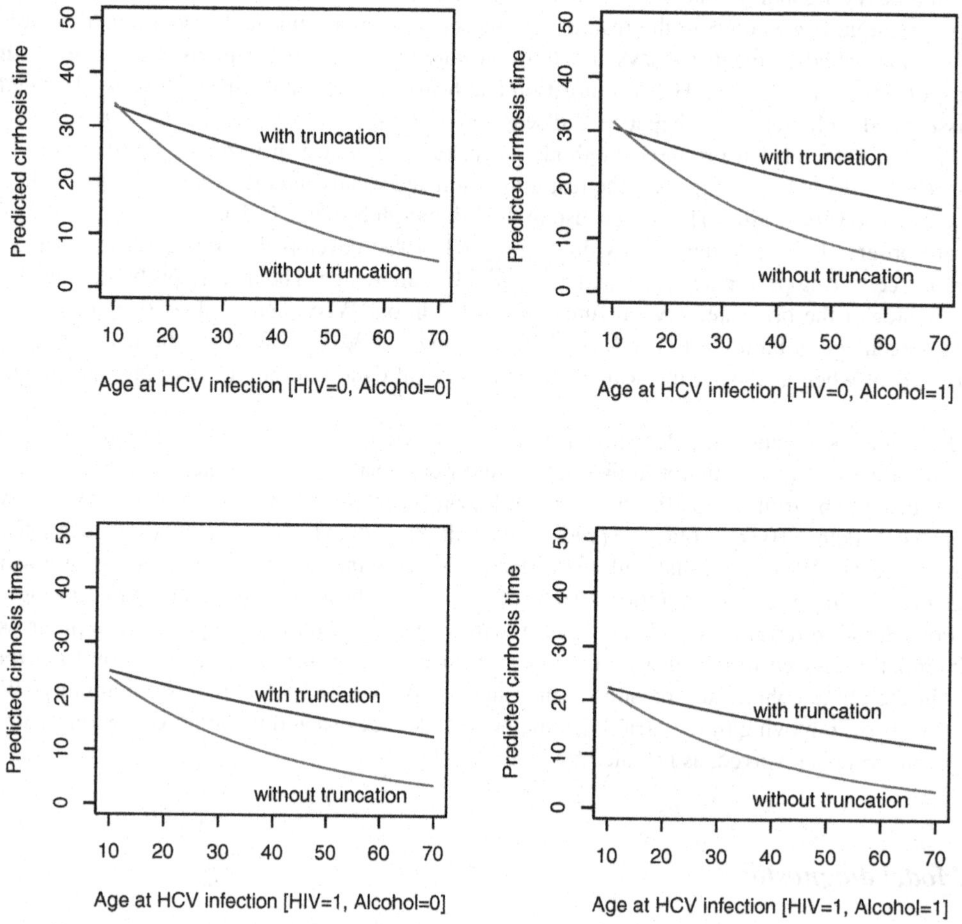

Fig. 4.1: Predictions of time from HCV infection to cirrhosis using accelerated failure time model with and without considering referral bias.

Then if the AFT model assumption is true, the revised test statistic in (4.71) should asymptotically follow a χ^2 distribution with degree of freedom $k - s - 1$, where s is the number of parameters to be estimated in the AFT model. Here s is taken to be 4 since there are 3 covariates and 1 intercept in the model.

Following this method, the calculated test statistic is 11.68 with a degree of freedom 7 ($k = 12$), which is less than the threshold value 14.07. Therefore we do not have significant evidence to reject the null hypothesis. Thus we conclude that the AFT model assumption here for the cirrhosis data is reasonable.

Chapter 5
Recent advances for truncated survival data

5.1 Linear transformation models

In practice, we may consider a class of transformation models, under which an unspecified or specified transformation of the survival time is linearly related to the covariates with various completely specified error distributions. Inference procedure can be derived from a class of generalised estimating equations for semiparametric models or from likelihood construction for parametric models. These transformation models provide useful alternatives to the nonparametric accelerated failure time model which has been discussed in Chapter 4.

5.1.1 *Semi-parametric models*

The semiparametric linear transformation models are given by

$$h(T) = -\boldsymbol{W}\beta + \varepsilon, \tag{5.1}$$

where T is time to the event of interest, $h(\cdot)$ is a completely unspecified strictly increasing function, and ε is a random error with distribution function F_ε. If F_ε is taken to be an extreme value distribution where $F_\varepsilon(t) = 1 - \exp\{-\exp(t)\}$, (5.1) gives the Cox proportional hazard model (Cox, 1972), while if F_ε is the standard logistic distribution function, (5.1) is the proportional odds model (Bennett, 1983; Dabrowska and Doksum, 1988; Pettitt, 1984; Murphy et al., 1997).

For survival data with only right censoring, Cheng et al. (1995) assumed a completely specified F_ε but unspecified $h(\cdot)$, and proposed a class of simple estimating equations for β in (5.1). Specifically, let T denote the failure time and let C denote the censoring time with a distribution function G. For the ith subject, one can observe $(X_i, \delta_i, \boldsymbol{W}_i)$, where $X_i = \min(T_i, C_i)$ and $\delta_i = I[T_i \leqslant C_i]$. They considered the case that

$$E\left\{\frac{\delta_j I[X_i \geqslant X_j]}{G^2(X_j)} - \theta(\boldsymbol{W}_{ij}\beta)\middle|\boldsymbol{W}_i, \boldsymbol{W}_j\right\} = 0, \tag{5.2}$$

where $\boldsymbol{W}_{ij} = \boldsymbol{W}_i - \boldsymbol{W}_j$, $i, j = 1, \ldots, n$, $i \neq j$, and

$$\theta(t) = \int_{-\infty}^{\infty} [1 - F_\varepsilon(t+s)] F_\varepsilon(ds). \tag{5.3}$$

They proposed an estimating equation

$$U(\beta) = \sum_{i=1}^{n}\sum_{j=1}^{n} \omega(\boldsymbol{W}_{ij}\beta)\boldsymbol{W}_{ij}\left\{\frac{\delta_j I[X_i \geqslant X_j]}{\hat{G}^2(X_j)} - \theta(\boldsymbol{W}_{ij}\beta)\right\} = 0, \tag{5.4}$$

Analysis for Time-to-Event Data under Censoring and Truncation.
http://dx.doi.org/10.1016/B978-0-12-805480-2.50005-8,

where $\omega(\cdot)$ is a weight function, and $\hat{G}$ is the Kaplan-Meier estimator for G. Their method was further developed in Cheng et al. (1997); Fine et al. (1998); Cai et al. (2000); Chen et al. (2002).

The method proposed by Cheng et al. (1995) can be extended to handle bivariate survival data with both truncation and censoring. Consider the cirrhosis data described in Example 1.5, where the two event times (R,T) are the time from HCV infection to referral to the clinic cohort and the time from HCV infection to the development of cirrhosis, respectively. Using the same notations in section 4.3, we have

$$E\left\{\frac{\delta_j I[X_i \geqslant X_j]}{G(R_i-,X_j-)G(R_j-,X_j-)} - \frac{\theta(\boldsymbol{W}_{ij}\boldsymbol{\beta})}{G(R_i-,0)G(R_j-,0)}\right\} = 0. \tag{5.5}$$

If the bivariate survival function G is known, we have the following unbiased estimating equation for $\boldsymbol{\beta}$ in (5.1),

$$U(\boldsymbol{\beta}) = \sum_{i=1}^{n}\sum_{j=1}^{n} \boldsymbol{W}_{ij}\left\{\frac{\delta_j I[X_i \geqslant X_j]}{G(R_i-,X_j-)G(R_j-,X_j-)} - \frac{\theta(\boldsymbol{W}_{ij}\boldsymbol{\beta})}{G(R_i-,0)G(R_j-,0)}\right\} = \boldsymbol{0}. \tag{5.6}$$

When G is unknown, it can be replaced by the consistent estimator discussed in section 3.5. Then we can use iterative algorithms to get a consistent regression parameter estimate $\hat{\boldsymbol{\beta}}$, which satisfies the above estimating equation.

5.1.2 Parametric models

Besides the semiparametric models discussed above, we may also consider a parametric model with completely specified transformation function $h(\cdot)$ and distribution of ε. For example, taken $h(\cdot) = \log(\cdot)$, the model in (5.1) can be written as

$$\log T = \boldsymbol{W}\boldsymbol{\beta} + \sigma\varepsilon, \tag{5.7}$$

where σ is a scale parameter and the distribution function of ε is still denoted by F_ε. Given the vector of covariates $\boldsymbol{W}$, the survival function of T (based on the AFT model in (4.1)) is

$$S(t|\boldsymbol{W}) = \bar{F}_\varepsilon\left(\frac{\log t - \boldsymbol{W}\boldsymbol{\beta}}{\sigma}\right),$$

where $\bar{F}_\varepsilon(\cdot) = 1 - F_\varepsilon(\cdot)$. For such parametric models, we may choose F_ε as normal distribution (then T is log-normal) or the standard extreme value distribution.

The maximum likelihood method can be used to estimate the parameters. In the presence of only right censoring, denote $(X_i, \delta_i, \boldsymbol{W}_i)$ as the observed data for the ith subject, where $X_i = \min\{\log T_i, \log C_i\}$ and $\delta_i = I[T_i \leq C_i]$. The likelihood function is given by

$$L(\boldsymbol{\beta},\sigma) \propto \prod_{i=1}^{n}\left[\frac{1}{\sigma}f\left(\frac{X_i - \boldsymbol{W}_i\boldsymbol{\beta}}{\sigma}\right)\right]^{\delta_i}\left[\bar{F}_\varepsilon\left(\frac{X_i - \boldsymbol{W}_i\boldsymbol{\beta}}{\sigma}\right)\right]^{1-\delta_i}. \tag{5.8}$$

Similarly, in the presence of only left censoring, the censoring indicator $\delta_i = I[T_i \geq C_i]$. Then the likelihood function is

$$L(\boldsymbol{\beta},\sigma) \propto \prod_{i=1}^{n}\left[\frac{1}{\sigma}f\left(\frac{X_i - \boldsymbol{W}_i\boldsymbol{\beta}}{\sigma}\right)\right]^{\delta_i}\left[F_\varepsilon\left(\frac{X_i - \boldsymbol{W}_i\boldsymbol{\beta}}{\sigma}\right)\right]^{1-\delta_i}. \tag{5.9}$$

In the presence of left truncation and right censoring, the observed data for the ith subject is denoted by $(X_i, Y_i, \delta_i, \boldsymbol{W}_i)$, where the truncation condition is $X > Y$. The likelihood function is then given by

$$L(\boldsymbol{\beta},\sigma) \propto \prod_{i=1}^{n}\left[\frac{1}{\sigma}f\left(\frac{X_i - \boldsymbol{W}_i\boldsymbol{\beta}}{\sigma}\right)\Big/\bar{F}_\varepsilon\left(\frac{Y_i - \boldsymbol{W}_i\boldsymbol{\beta}}{\sigma}\right)\right]^{\delta_i}\left[\bar{F}_\varepsilon\left(\frac{X_i - \boldsymbol{W}_i\boldsymbol{\beta}}{\sigma}\right)\Big/\bar{F}_\varepsilon\left(\frac{Y_i - \boldsymbol{W}_i\boldsymbol{\beta}}{\sigma}\right)\right]^{1-\delta_i}. \tag{5.10}$$

Similarly, for right truncated and right censored data under the truncation condition $X < Y$, the likelihood function is

$$L(\beta,\sigma) \propto \prod_{i=1}^{n}\left[\frac{1}{\sigma}f\left(\frac{X_i - \boldsymbol{W}_i\beta}{\sigma}\right)\Big/F_{\varepsilon}\left(\frac{Y_i - \boldsymbol{W}_i\beta}{\sigma}\right)\right]^{\delta_i}\left[\bar{F}_{\varepsilon}\left(\frac{X_i - \boldsymbol{W}_i\beta}{\sigma}\right)\Big/F_{\varepsilon}\left(\frac{Y_i - \boldsymbol{W}_i\beta}{\sigma}\right)\right]^{1-\delta_i}. \tag{5.11}$$

Then the maximum likelihood estimates for the regression parameter vector β and the scale parameter σ can be found by solving the following estimating equations

$$\frac{\partial \log L(\beta,\sigma)}{\partial \beta} = 0,$$
$$\frac{\partial \log L(\beta,\sigma)}{\partial \sigma} = 0,$$

using iterative algorithms such as the Newton-Raphson method.

5.2 Joint modelling of survival events and longitudinal data under random truncation

Another active research area is the joint modelling of truncated survival events with longitudinal data. In longitudinal studies repeated measurements from each subject are collected and these repeated measurements usually correspond to a subject's biomarker. For example HIV patients may be followed over time and repeated measurements such as CD4 counts are collected to characterize patients' immune status (the biomarker). Random errors always occur when collecting experimental measurements. Therefore, the real biomarker can only be observed at certain predetermined time points (the times to collect longitudinal data) and with random errors. In this section we will study how to analyse the relation between the survival events and the latent biomarker, with the occurrence of random truncation.

5.2.1 *General introduction of Joint models*

In Chapter 2, we have discussed the proportional hazard regression models, with the covariate $\boldsymbol{Z}$ fully observed. However, when the covariate is time-dependent and corresponds to a biomarker of a subject, we need to measure it repeatedly. The measurement may be even subject to random errors. In particular, we may denote such continuous covariate random process as $\mu_i(t)$, which can only be observed intermittently at times $\boldsymbol{t}_i = (t_{i1},\cdots,t_{i,m_i})$ with random error. Then the observed longitudinal data are given by

$$Y_i^{(j)} = \mu_i(t_{ij}) + \varepsilon_i^{(j)}, \quad j = 1,\cdots,m_i; i = 1,\cdots,n, \tag{5.12}$$

where the error term $\{\varepsilon_i^{(j)}, j = 1,\cdots,m_i, i = 1,\cdots,n\}$ are i.i.d. and normally distributed with mean 0 and variance σ^2. The above model can be released to more general scenarios. For example εs are i.i.d. and have a distribution function F_ε with mean 0 or $\varepsilon^{(j)}$s have a serial correlation. Although in this section we assume that $\boldsymbol{t} = (t_1,\cdots,t_{m_i})$ are predetermined fixed time points, these time points $\boldsymbol{t}$ can also be treated as random. Readers may refer to Diggle et al. (2002) for more details of longitudinal studies.

Example 5.1. The primary biliary cirrhosis (PBC) study in Murtaugh et al. (1994). The PBC is a chronic, fatal, but rare liver disease characterized by inflammatory destruction of the small bile ducts within the liver, which eventually leads to cirrhosis of the liver. Patients often present abnormalities in their blood tests, such as elevated and gradually increased serum bilirubin. Patients in this study will have their blood tests (for example including measurements of serum bilirubin) at certain predetermined time points.

Therefore the relation of the observed measurements and the biomarker (real serum bilirubin level) can be modelled via (5.12).

In the above example, a subject may also experience a survival event (say death). The proportional hazard regression model for the survival time X is

$$\Lambda_X(dt) = \Lambda_0(dt)\exp(Z(t)\beta + \gamma\mu(t)), \tag{5.13}$$

where $Z(t)$ is a random vector of covariates (could be time dependent) and can be fully observed and β and γ are unknown parameters to be estimated. It is also assumed that ε (in model (5.12)) is independent of all other random variables involved in (5.12) and (5.13).

The unobserved biomarker process $\mu(t)$ can be modelled via many different approaches. For example, we may model $\mu(t)$ as a polynomial function of t, where the coefficients of the polynomial are interpreted as random effects (they are different for different subjects) (Tisiatis and Davidian, 2001). Sometimes, for simplicity, the model $\mu_i(t) = a_i\mu_0(t)$ with a single random effect a_i involved and a nonparametric function $\mu_0(t)$ (common to all subjects), for example see Ding and Wang (2008). There are even other types of approaches to *joint* the survival model and the longitudinal model. Readers may refer to Tisiatis and Davidian (2004) for a comprehensive review.

For the data described in Example 5.1, Fleming and Harrington (1991) provide the proportional hazard regression analysis results by using the cross-sectional data (the baseline longitudinal observations only), but it makes more sense to use longitudinal data. In this section, we will focus on the joint modelling of (5.12) and (5.13) mentioned above.

Longitudinal data are the most common type of data in areas such as epidemiology, clinical research and so on. It is very important to consider longitudinal observations in survival analysis, rather than using cross-sectional data. This is because the timing of disease onset is usually correlated with recent changes in patient exposure. Longitudinal data can have the records of changes over a long period. Another key strength of using longitudinal data is the ability to measure change in outcomes and/or exposure at the individual level. The benefits of such joint modelling approach are not without a cost. The challenge of the joint modelling of truncated survival events and longitudinal data lies in dealing with unobserved biomarker process and the truncation. We finish this subsection by presenting a real example and then introduce the methodology in the following subsections.

Example 5.2. Su and Wang (2012) considered an analysis on the data from a multi-centre HIV study in Italy, which was originally analysed by Rezza et al. (1989). The primary event of interest is the incubation period of acquired immunodeciency syndrome (AIDS), i.e. time from detection of HIV-infection until the onset of AIDS. Some patients received the HAART treatment at various times, which gives a fully observable covariate process. In addition, the second covariate, the CD4 counts, that are observed only intermittently at follow-up visits, result in longitudinal measurements. The main research interest is to determine the HAART treatment effects on reducing the risk of developing AIDS and analyse how the incubation period of AIDS depends on the CD4 T-cell counts in HIV infected subjects.

One feature of the data is that the event time, the incubation period, is subject to left truncation and right censoring, since patients were recruited to the study at various times after the study began. Therefore, only patients who have not developed AIDS at the time of recruitment are included in the study.

5.2.2 Methodology

As we discussed in Chapter 2, the survival event X is subject to random censoring, with notations $\tilde{X} = \min\{X, C\}$, $\delta = I[X \le C]$, and random truncation, i.e. we can only observe the subject information if $\tilde{X} > L$. The observed survival data are denoted as $(\tilde{X}_i, L_i, \delta_i)$. Using the same notations as those in Chapter 2, we define the counting process $N_i(t) = I[\tilde{X}_i \le t, \delta_i = 1]$ and at risk process $H_i(t) = I[\tilde{X}_i \ge t > L_i]$.

Define $\mathscr{Z}_{i,t} = \sigma\{Z_i(s); 0 \le s \le t\}$, $\mathscr{G}_{i,t} = \sigma\{\mu_i(t); 0 \le s \le t\}$, $\mathscr{H}_{i,t} = \sigma\{Y_{i,j}; \forall j,$ such that $t_{i,j} \le t\}$ and $\mathscr{F}_{i,t} = \sigma\{N_i(u), N_i^C(u), I[L_i \le u] : 0 \le u \le t\}$. The filtration $\mathscr{Z}_{i,t}$ means the information for the fully observed covariates, the filtration $\mathscr{F}_{i,t}$ means the information from the survival events and the filtration $\mathscr{H}_{i,t}$

means the longitudinal data information collected by time t. Note that the filtration $\mathscr{G}_{i,t}$, the information of real biomarker process $\mu_i(t)$, is actually not observed. Following the notations in early chapters, letters with a subscript i, $(\tilde{X}_i, L_i, \delta_i, \mathscr{Z}_{i,t}, \mathscr{G}_{i,t}, \mathscr{H}_{i,t})$, denote the information for the ith observed subject. Note that $\mathscr{G}_{i,t}$ is unknown. All subjects in the whole population are denoted as $(\tilde{X}_i^*, L_i^*, \delta_i^*, \mathscr{Z}_{i,t}^*, \mathscr{G}_{i,t}^*, \mathscr{H}_{i,t}^*), i = 1, 2, \cdots$. The ith subject in the biased data corresponds to the k_ith observation in the population. Note that we may also denote a typical subject in the population as $(\tilde{X}, L, \delta, \mathscr{Z}_t, \mathscr{G}_t, \mathscr{H}_t)$.

The theory introduced in Chapter 2 will not be valid here since the process $\mu_i(t)$ (or $\mathscr{G}_{i,t}$) can only be observed intermittently with errors. In this section, we briefly discuss how to extend the method in Chapter 2 for standard proportional hazard models to the cases with longitudinal data.

As we mentioned in section 5.1, the model for $\mu(t)$ plays a key role in the joint modelling approach. We here, however, do not employ any particular model for $\mu(t)$, but to deal with it in a very general approach. Suppose we can find a continuous process $\hat{\mu}_i(t)$, as an unbiased estimate for $\mu_i(t)$. A simple approach is to replace $\mu_i(t)$ in (5.13) by its unbiased estimate $\hat{\mu}_i(t)$ and employ the standard martingale method introduced in Chapter 2. However, such a naive approach will give biased estimate (Henderson et al., 2000). We may also use EM algorithms, for example Ding and Wang (2008). The EM algorithm, however, may not provide good estimate due to the (Monte Carlo) integration involved for a large dimensional random effects, as Ding and Wang (2008) pointed out. In this section, we introduce a simple approach based on martingale theories, which does not need intensive computations.

Assumption 5.2.1 *1. Suppose that we can find an unbiased estimate $\hat{\mu}_i(t)$ for $\mu_i(t)$, such that $\hat{\mu}_i(t)$ is predictable with respect to $\mathscr{H}_{i,t}$.*
2. *Conditional on $\mu_i(t)$, $\hat{\mu}_i(t)$ follows a normal distribution with mean $\mu_i(t)$ and variance $\sigma_i^2(t)$, which only depends on the distribution of ε and $\mathbf{t}_i$.*

Note that such an unbiased estimate, required in Assumption 5.2.1, is not difficult to find. For example, if we assume that $\mu_i(t)$ is a linear function of certain unknown random effects $(v_{i,0}, \cdots, v_{i,q})$, $\mu_i(t) = v_{i,0} + v_{i,1}t + \cdots, v_{i,q}t^q$, then $\hat{\mu}_i(t)$ is the predicted value based on the Least Squares estimate of $(v_{i,0}, v_{i,1}, \cdots, v_{i,q})$ using the observations $Y_i^{(j)}$, for all j such that $t_{i,j} \leq t$. For such a polynomial model of $\mu_i(t)$, part 2 of Assumption 5.2.1 is also true, if we assume ε is normally distributed.

We have the following lemma.

Lemma 5.1. *Under Assumption 5.2.1, we have*

$$\begin{aligned}
&\mathbf{P}(\tilde{X}_i \in dt, \delta_i = 1 | \mathscr{Z}_{i,t_-}, \mathscr{F}_{i,t_-}) \\
&= H_i(t)\Lambda_0(dt)\exp\left(\boldsymbol{Z}_i(t)\beta\right) E\left\{\exp(\gamma\mu_i(t)) | \mathscr{Z}_{i,t_-}, \mathscr{F}_{i,t_-}\right\} \\
&= H_i(t)\Lambda_0(dt)\exp\left(\boldsymbol{Z}_i(t)\beta - \frac{\gamma^2}{2}\sigma_i^2(t)\right) E\left\{E\left[\exp(\gamma\hat{\mu}_i(t)) | \mathscr{G}_{i,t_-}\right] | \mathscr{Z}_{i,t_-}, \mathscr{F}_{i,t_-}\right\}. \qquad (5.14)
\end{aligned}$$

Proof. First, conditional on $\mu_i(t)$, $\hat{\mu}_i(t)$ follows a normal distribution with mean $\mu_i(t)$ and variance $\sigma_i^2(t)$. This implies that

$$E\left[\exp\left(\gamma\hat{\mu}_i(t) - \mu_i(t)\right) | \mathscr{G}_{i,t_-}\right] = \exp\left(\frac{\gamma^2\sigma_i^2(t)}{2}\right)$$

or

$$E\left[\exp\left(\gamma\hat{\mu}_i(t) - \frac{\gamma^2\sigma_i^2(t)}{2}\right) | \mathscr{G}_{i,t_-}\right] = E\left[\exp\left(\mu_i(t)\right) | \mathscr{G}_{i,t_-}\right]. \qquad (5.15)$$

Therefore, we have

$$\begin{aligned}
&\mathbf{P}(\tilde{X}_i \in dt, \delta_i = 1 | \mathscr{Z}_{i,t-}, \mathscr{F}_{i,t-}) \\
&= E\left\{\mathbf{P}(\tilde{X}_i \in dt, \delta_i = 1 | \mathscr{Z}_{i,t-}, \mathscr{G}_{i,t-}, \mathscr{F}_{i,t-}) | \mathscr{Z}_{i,t-}, \mathscr{F}_{i,t-}\right\} \\
&= H_i(t)\Lambda_0(dt)\exp\left(Z_i(t)\beta\right) E\left[\exp(\gamma\mu_i(t)) | \mathscr{Z}_{i,t-}, \mathscr{F}_{i,t-}\right] \\
&= H_i(t)\Lambda_0(dt)\exp\left(Z_i(t)\beta - \frac{\gamma^2}{2}\sigma_i^2(t)\right) E\left\{E\left[\exp(\gamma\hat{\mu}(t)) | \mathscr{G}_{i,t-}\right] | \mathscr{Z}_{i,t-}, \mathscr{F}_{i,t-}\right\}. \qquad (5.16)
\end{aligned}$$

■

Apart from the information of $\mathscr{Z}$ and $\mathscr{F}$ we also have the longitudinal information $\mathscr{H}$, therefore similar to Lemma 5.1, we also have

$$\begin{aligned}
&E(dN_i(t) | \mathscr{Z}_{i,t-}, \mathscr{H}_{i,t-}, \mathscr{F}_{i,t-}) \\
&= H_i(t)\Lambda_0(dt)\exp\left(Z_i(t)\beta\right) E\left\{\exp(\gamma\mu_i(t)) | \mathscr{Z}_{i,t-}, \mathscr{H}_{i,t-}, \mathscr{F}_{i,t-}\right\}. \qquad (5.17)
\end{aligned}$$

Note that it may not be easy to work out $E\left\{\exp(\gamma\mu_i(t)) | \mathscr{Z}_{i,t-}, \mathscr{H}_{i,t-}, \mathscr{F}_{i,t-}\right\}$ explicitly. However, by noticing (5.15), we know that

$$E\left[\exp\left(\gamma\hat{\mu}_i(t) - \frac{\gamma^2\sigma_i^2(t)}{2}\right)\right] = E\left[\exp\left(\mu_i(t)\right)\right]. \qquad (5.18)$$

Now we present a heuristic argument first and then explain why it is correct. Equation (5.18) and the fact that $\exp\left(\gamma\hat{\mu}_i(t) - \frac{\gamma^2\sigma_i^2(t)}{2}\right)$ is measurable with respect to $\mathscr{H}_{i,t-}$, imply that we may use $\exp\left(\gamma\hat{\mu}_i(t) - \frac{\gamma^2\sigma_i^2(t)}{2}\right)$ to replace $E\left\{\exp(\gamma\mu_i(t)) | \mathscr{Z}_{i,t-}, \mathscr{H}_{i,t-}, \mathscr{F}_{i,t-}\right\}$ in equation (5.17), i.e.

$$\begin{aligned}
&E(dN_i(t) | \mathscr{Z}_{i,t-}, \mathscr{H}_{i,t-}, \mathscr{F}_{i,t-}) \\
&= H_i(t)\Lambda_0(dt)\exp\left(Z_i(t)\beta + \gamma\hat{\mu}_i(t) - \frac{\gamma^2}{2}\sigma_i^2(t)\right) \\
&:= \tilde{A}_i(t; \beta, \gamma, \Lambda_0). \qquad (5.19)
\end{aligned}$$

Note that equation (5.19) is true if we consider the revised proportional hazards model

$$\Lambda_X(dt) = \Lambda_0(dt)\exp\left(Z(t)\beta + \gamma\hat{\mu}(t) - \frac{\gamma^2}{2}\sigma^2(t)\right), \qquad (5.20)$$

where $\sigma^2(t)$ is the variance of $\hat{\mu}(t)$ given $\mu(t)$. On the other hand, because of equation (5.18), model (5.20) is an unbiased version of (5.13), i.e. if $(\Lambda_0^*, \beta^*, \gamma^*)$ is the true parameter for model (5.20) then it is also the true parameter for model (5.13). Therefore, it makes sense to consider model (5.20) and the martingale, with respect to the filtration $\mathscr{Z}_{i,t} \vee \mathscr{H}_{i,t} \vee \mathscr{F}_{i,t}$,

$$M_i(dt) = dN_i(t) - H_i(t)\Lambda_0(dt)\exp\left(Z_i(t)\beta + \gamma\hat{\mu}_i(t) - \frac{\gamma^2}{2}\sigma_i^2(t)\right). \qquad (5.21)$$

We can use equation (5.19) to construct the log-likelihood and score function, via a similar approach as that in Chapter 2.

5.2.3 *The likelihood function and the estimate*

Suppose that we consider the time period $[0, \tau]$. Using equation (5.19) and similar arguments as that in Chapter 2, the log partial likelihood is given by

$$\tilde{l}(\beta,\gamma) = \int_{[0,\tau]} \left[Z_i(t)\beta + \gamma\hat{\mu}_i(t) - \frac{\gamma^2}{2}\sigma_i^2(t) - \log(\tilde{S}^{(0)}(\beta,\gamma,t)) \right] dN_i(t) \tag{5.22}$$

where

$$\tilde{S}^{(0)}(\beta,\gamma,t) = n^{-1}\sum_i H_i(t)\exp\left(Z_i(t)\beta + \gamma\hat{\mu}_i(t) - \frac{\gamma^2}{2}\sigma_i^2(t) \right). \tag{5.23}$$

The score function is

$$\tilde{U}(\beta,\gamma,\tau) = \sum_{i=1}^{n} \int_{[0,\tau]} \left[\begin{pmatrix} Z_i(t) \\ \hat{\mu}_i(t) - \gamma\sigma_i^2(t) \end{pmatrix} - \frac{\tilde{S}^{(1)}(\beta,\gamma,t)}{\tilde{S}^{(0)}(\beta,\gamma,t)} \right] dN_i(t) \tag{5.24}$$

where $\tilde{S}^{(1)}(\beta,\gamma,t)$ is the first-order derivative of $\tilde{S}^{(0)}(\beta,\gamma,t)$ with respect to the parameters,

$$\tilde{S}^{(1)}(\beta,\gamma,t) = n^{-1}\sum_i H_i(t)\begin{pmatrix} Z_i(t) \\ \hat{\mu}_i(t) - \gamma\sigma_i^2(t) \end{pmatrix}\exp\left(Z_i(t)\beta + \gamma\hat{\mu}_i(t) - \frac{\gamma^2}{2}\sigma_i^2(t) \right). \tag{5.25}$$

We may also need the second-order derivative of $\tilde{S}(\beta,\gamma,t)$, given by

$$\tilde{S}^{(2)}(\beta,\gamma,t) = n^{-1}\sum_i H_i(t)\left[\begin{pmatrix} Z_i(t) \\ \hat{\mu}_i(t) - \gamma\sigma_i^2(t) \end{pmatrix}^{\otimes 2} - \begin{pmatrix} 0 & 0 \\ 0 & \sigma_i^2(t) \end{pmatrix} \right]\exp\left(Z_i(t)\beta + \gamma\hat{\mu}_i(t) - \frac{\gamma^2}{2}\sigma_i^2(t) \right). \tag{5.26}$$

The solution of the score function gives us the estimate $(\hat{\beta},\hat{\gamma})$, which can be found via Newton-Rhapson algorithm. The estimate for Λ_0 is given by

$$\hat{\Lambda}_0(dt) = \frac{n^{-1}\sum_{i=1}^{n} dN_i(t)}{\tilde{S}^{(0)}(\hat{\beta},\hat{\gamma},t)}. \tag{5.27}$$

Note that if $\mu_i(t)$ is given (or $\mathscr{G}_i$ is available), we have

$$E(dN_i(t)|\mathscr{Z}_{i,t_-},\mathscr{G}_{i,t_-},\mathscr{F}_{i,t_-}) = H_i(t)\Lambda_0(dt)\exp(Z_i(t)\beta + \gamma\mu_i(t)) \tag{5.28}$$

and the score function

$$U(\beta,\gamma,\tau) = \sum_{i=1}^{n} \int_{[0,\tau]} \left[\begin{pmatrix} Z_i(t) \\ \mu_i(t) \end{pmatrix} - \frac{S^{(1)}(\beta,\gamma,t)}{S^{(0)}(\beta,\gamma,t)} \right] dN_i(t) \tag{5.29}$$

where

$$\begin{aligned} S^{(1)}(\beta,\gamma,t) &= n^{-1}\sum_i H_i(t)\begin{pmatrix} Z_i(t) \\ \mu_i(t) \end{pmatrix}\exp(Z_i(t)\beta + \gamma\mu_i(t)) \\ S^{(0)}(\beta,\gamma,t) &= n^{-1}\sum_i H_i(t)\exp(Z_i(t)\beta + \gamma\mu_i(t)) \end{aligned} \tag{5.30}$$

We can also define

$$S^{(2)}(\beta,\gamma,t) = n^{-1}\sum_i H_i(t)\begin{pmatrix} Z_i(t) \\ \mu_i(t) \end{pmatrix}^{\otimes 2}\exp(Z_i(t)\beta + \gamma\mu_i(t)). \tag{5.31}$$

If we use $\tilde{S}^{(k)}, S^{(k)}, k = 0,1,2$ to denote the above scalar, vector and matrix functions, respectively, then clearly $\lim_{n\to\infty} S^{(k)}(\beta,\gamma,t) = \lim_{n\to\infty} \tilde{S}^{(k)}(\beta,\gamma,t) := s^{(k)}(\beta,\gamma,t)$.

Denote $(\beta_0, \gamma_0, \Lambda_{0,0})$ as the true parameter values. Under conditions similar to Condition 2.4.2, we also have the asymptotic distributions for the parameter estimates, i.e. $\sqrt{n}((\hat{\beta},\hat{\gamma})' - (\beta_0,\gamma_0))$ is asymptotically normal with mean $\mathbf{0}$ and covaraince matrix $\boldsymbol{v}(\beta_0, \gamma_0, \Lambda_{0,0}, \tau)^{-1}$,

$$\boldsymbol{v}(\beta,\gamma,t) = \int_{[0,t]} \left[\boldsymbol{s}^{(2)}(\beta,\gamma,s) - \frac{\boldsymbol{s}^{(1)}(\beta,\gamma,s)^{\otimes 2}}{s^{(0)}(\beta,\gamma,s)} \right] d\Lambda_{0,0}(s), \tag{5.32}$$

and $\sqrt{n}(\hat{\Lambda}_0(t) - \Lambda_{0,0}(t))$ is asymptotically normal with mean 0 and variance

$$\sigma_\Lambda^2(t) = \int_{[0,t]} \frac{d\Lambda_{0,0}(s)}{s^{(0)}(\beta,\gamma,s)} + \boldsymbol{q}(\beta,\gamma,t)'\boldsymbol{v}(\beta,\gamma,t)^{-1}\boldsymbol{q}(\beta,\gamma,t), \tag{5.33}$$

where

$$\boldsymbol{q}(\beta,\gamma,t) = \int_{[0,t]} \frac{\boldsymbol{s}^{(1)}(\beta,\gamma,s)}{s^{(0)}(\beta,\gamma,s)} d\Lambda_{0,0}(s). \tag{5.34}$$

An estimate of $\boldsymbol{v}(\beta,\gamma,t)$ is given by $\tilde{V}(\hat{\beta},\hat{\gamma},t)$,

$$\tilde{V}(\beta,\gamma,t) = n^{-1}\sum_{i=1}^{n} \int_{[0,\tau]} \left[\frac{\tilde{\boldsymbol{S}}^{(2)}(\hat{\beta},\hat{\gamma},s)}{\tilde{S}^{(0)}(\hat{\beta},\hat{\gamma},s)} - \frac{\tilde{\boldsymbol{S}}^{(1)}(\hat{\beta},\hat{\gamma},s)^{\otimes 2}}{\tilde{S}^{(0)}(\hat{\beta},\hat{\gamma},s)^2} \right] dN_i(s), \tag{5.35}$$

and an estimate of $\sigma_\Lambda^2(t)$ is given by

$$\hat{\sigma}_\Lambda^2(t) = \int_{[0,t]} \frac{d\hat{\Lambda}_0(s)}{\tilde{S}^{(0)}(\hat{\beta},\hat{\gamma},s)} + \boldsymbol{Q}(\hat{\beta},\hat{\gamma},t)'\tilde{\boldsymbol{V}}(\hat{\beta},\hat{\gamma},t)^{-1}\boldsymbol{Q}(\hat{\beta},\hat{\gamma},t),$$
$$\boldsymbol{Q}(\beta,\gamma,t) = n^{-1}\sum_i \int_{[0,t]} \frac{\tilde{\boldsymbol{S}}^{(1)}(\beta,\gamma,s)}{\tilde{S}^{(0)}(\beta,\gamma,s)^2} dN_i(s). \tag{5.36}$$

Note that all the estimators given above assume that σ^2 (the variance of ε) is unknown. In practice, we can use its consistent estimator $\hat{\sigma}^2$. For example, if we assume the polynomial model $\mu_i(t) = \nu_{i,0} + \nu_{i,1}t + \cdots, \nu_{i,q}t^q$ and $\hat{\mu}_i(t)$ is the predictor based on Least Squares estimates, then $\hat{\sigma}^2 = n^{-1}\sum_i (m_i - 1)^{-1}\sum_{j=1}^{m_i}(Y_i^{(j)} - \hat{\mu}_i(t_{ij}))^2$.

5.2.4 More general model assumptions on ε

The method presented in this section does not require any particular distribution assumption (or even modelling assumption) on μ_i, but does assume that ε_i are independent and identically distributed, as a normal distribution. If ε is not normally distributed, the above method will not work since equation (5.14) (or Lemma 5.1) is not valid.

If we define

$$\psi_i^{(k)}(\gamma,s) = E\left[(\hat{\mu}_i(s) - \mu_i(s))^k \exp(\gamma(\hat{\mu}_i(s) - \mu_i(s))) | \mathscr{G}_{i,s} \right], \tag{5.37}$$

we may be able to estimate $\psi_i^{(k)}$ if the distribution F_ε is specified. Then we will have a more general version of the method introduced in the previous subsections. Note that $\psi_i^{(k)}$ is the kth derivative of $\psi_i^{(0)}$ with respect to γ.

Now if we consider the alternative model

$$\Lambda_X(dt) = \Lambda_0(dt)\exp\left(Z_i(t)\beta + \gamma\hat{\mu}_i(t)\right)\frac{1}{\psi_i^{(0)}(\gamma,s)}, \tag{5.38}$$

we know that if $\beta_0, \gamma_0, \Lambda_{0,0}$ are the true parameters for (5.38), they should also be the true parameters for (5.13) since

$$\frac{\exp(Z_i(s)\beta + \gamma\hat{\mu}_i(s))}{\psi^{(0)}(\gamma,s)} = \exp(Z_i(s)\beta + \gamma\mu_i(s))\frac{\exp(\gamma(\hat{\mu}_i(s) - \mu_i(s)))}{\psi_i^{(0)}(\gamma,s)},$$
$$E\left[\frac{\exp(Z_i(s)\beta + \gamma\hat{\mu}_i(s))}{\psi_i^{(0)}(\gamma,s)}|\mathscr{G}_{i,s}\right] = \exp(Z_i(s)\beta + \gamma\mu_i(s)). \tag{5.39}$$

Under model (5.38), we have

$$dM_i(t) := dN_i(t) - H_i(t)\frac{\exp(Z_i(t)\beta + \gamma\hat{\mu}_i(t))}{\psi_i^{(0)}(\gamma,t)}d\Lambda_0(t) \tag{5.40}$$

is a martingale with respect to the filtration $\mathscr{F}_{i,t} \vee \mathscr{H}_{i,t} \vee \mathscr{Z}_{i,t}$.

With similar arguments in the previous section, we have the log partial likelihood given by

$$\int_{[0,\tau]}\left\{Z_i(s)\beta + \gamma\hat{\mu}_i(s) - \log\psi_i^{(0)}(\gamma,s) - \log\left[\check{S}(\beta,\gamma,s)\right]\right\}dN_i(t), \tag{5.41}$$

where

$$\check{S}^{(0)}(\beta,\gamma,s) = n^{-1}\sum_{i=1}^{n}\exp\left(Z_i(t)\beta + \gamma\hat{\mu}_i(t)\right)\frac{1}{\psi_i^{(0)}(\gamma,s)}. \tag{5.42}$$

Then the score function is given by

$$\check{U}(\beta,\gamma,\tau) = \int_{[0,\tau]}\left\{\begin{pmatrix} Z_i(s) \\ \hat{\mu}_i(s) - \frac{\psi_i^{(1)}(\gamma,s)}{\psi_i^{(0)}(\gamma,s)}\end{pmatrix} - \frac{\check{S}^{(1)}(\beta,\gamma,s)}{\check{S}^{(0)}(\beta,\gamma,s)}\right\}dN_i(t), \tag{5.43}$$

where

$$\check{S}^{(1)}(\beta,\gamma,s) = n^{-1}\sum_{i=1}^{n}\begin{pmatrix} Z_i(s) \\ \hat{\mu}_i(s) - \frac{\psi_i^{(1)}(\gamma,s)}{\psi_i^{(0)}(\gamma,s)}\end{pmatrix}\exp\left(Z_i(t)\beta + \gamma\hat{\mu}_i(t)\right)\frac{1}{\psi_i^{(0)}(\gamma,s)}. \tag{5.44}$$

And also we may need to define

$$\check{S}^{(2)}(\beta,\gamma,s) = n^{-1}\sum_{i=1}^{n}\left[\begin{pmatrix} Z_i(s) \\ \hat{\mu}_i(s) - \frac{\psi_i^{(1)}(\gamma,s)}{\psi_i^{(0)}(\gamma,s)}\end{pmatrix}^{\otimes 2} - \left(\frac{\psi_i^{(2)}(\gamma,s)}{\psi_i^{(0)}(\gamma,s)} - \frac{\psi_i^{(1)}(\gamma,s)^2}{\psi_i^{(0)}(\gamma,s)^2}\right)\right]\exp\left(Z_i(t)\beta + \gamma\hat{\mu}_i(t)\right)\frac{1}{\psi_i^{(0)}(\gamma,s)}. \tag{5.45}$$

In practice, $\varepsilon_i^{(1)}, \cdots, \varepsilon_i^{(m_i)}$ may be correlated, with a multinormal distribution $N(\mathbf{0}, \Sigma_i)$. We need to specify a particular model assumption on Σ_i, to avoid overfitting. For example, we may assume that $\varepsilon_i^{(j)}$ is made of a serial correlation and a measurement error, i.e.

$$\varepsilon_i^{(j)} = \xi_{i,j} + \zeta_{i,j}$$

where $\zeta_{i,j}, j = 1, \cdots, m_i$ are i.i.d. but $\xi_{i,j}, j = 1, \cdots, m_i$ are correlated with a covariance matrix $\Gamma_i = \{a_{i,j}\}$. The elements $a_{i,j} = \rho|t_i - t_j|$. One may also use a more general modelling approach, the mean-covariance modelling method in C. Leng and Pan (2010) to estimate the covariance matrix for the longitudinal observations Y. In practice, modelling the covariance matrix of Y may be more important in some studies, where the hazard rate of survival events depends on the variance of the biomarker (for example some disease could depends on the variation of the difference of systolic blood pressure and diastolic pressure).

5.2.5 Discussion on the bivariate case

In practice, the truncation may be because of truncation on another variable, say T, not the survival time X. For example, in a typical hepatitis C cohort study, each subject could experience a referral time T, a cirrhosis time X and a number of repeated longitudinal measurements $Y^{(j)}$ and some covariates $\boldsymbol{Z}$. The subject information is subject to random (say right) truncation, $T \leq L$, i.e. only patients who referred to hospital before a certain time point will be recorded.

Things become more complicated for such bivariate cases. If the hazard rate of X does not depend on the referral time, then we may use the same model as that in section 5.1.1 to analyse the data but use the truncation condition $T \leq L$ to debias the analysis result. Note that although X does not depend on T (in terms of a causal relationship), they may still be correlated (for example because they are both related to a common factor). If the hazard rate of X does depend on the referral time (in terms of a causal relationship), we can consider the following model for the survival part,

$$\Lambda_X(dt) = \Lambda_0(dt)\exp\left(\boldsymbol{Z}(t)\beta + \gamma\mu_i(t) + \eta f(T)\right)$$

which f a function which specifies how Λ_X and T are related. This is left to future work.

Bibliography

Akritas, M. G. and Keilegom, I. V. (2003). Estimation of bivariate and marginal distributions with censored data. *Journal of Royal Statistical Society*, 65:457–471.

Andersen, P. K., Borgan, r., Gill, R. D., and Keiding, N. (1993). *Statistical Models Based on Counting Processes*. Springer.

Bao, Y., He, S., and Mei, C. (2007). The Koul-Susarla-van Ryzin and weighted least squares estimates for censored linear regression model: A comparative study. *Computational Statistics & Data Analysis*, 51:6488–6497.

Bennett, S. (1983). Analysis of survival data by the proportional odds model. *Statistics in Medicine*, 2:273–277.

Buckley, J. and James, I. (1979). Linear regression with censored data. *Biometrika*, 66:429–436.

Burke, M. D. (1988). Estimation of a bivariate distribution function under random censorship. *Biometrika*, 75:379–382.

C. Leng, W. Z. and Pan, J. (2010). Semiparametric mean-covariance regression analysis for longitudinal data. *Journal of the American Statistical Association*, 105:181–193.

Cai, T., Wei, L. J., and Wilcox, M. (2000). Semiparametric regression analysis for clusterd failure time data. *Biometrika*, 87(4):867 – 878.

Campbell, G. (1981). Nonparametric bivariate estimation with randomly censored data. *Biometrika*, 68:417–422.

Chen, K., Jin, Z., and Ying, Z. (2002). Semiparametric analysis of transformation models with censored data. *Biometrika*, 89:659–668.

Chen, Y.-I. and Wolfe, D. A. (2000). Umbrella tests for right-censored survival data. *Statistica Sinica*, 10:595–612.

Cheng, S., Wei, L., and Ying, Z. (1995). Analysis of transformation models with censored data. *Biometrika*, 82:659–668.

Cheng, S., Wei, L., and Ying, Z. (1997). Predicting survival probabilities with semiparametric transformation models. *Journal of the American Statistical Association*, 92:227–235.

Clayton, D. (1978). A model for association in bivariate life tables and its application in epidemiology studies of familial tendency in chronic disease incidence. *Biometrika*, 65:141–151.

Cox, D. (1972). Regression models and life-tables. *Journal of Royal Statistical Society*, 34:187–202.

Cox, D. R. and Oakes, D. (1984). *Analysis of Survival Data*. Chapman and Hall, London.

Dabrowska, D. (1988). Kaplan-Meier estimate on the plane. *The Annals of Statistics*, 16:1475–1489.

Dabrowska, D. (1989). Kaplan-Meier estimate on the plane: weak convergence, lil, and the bootstrap. *Journal of Multivariate Analysis*, 29:308–325.

Dabrowska, D. and Doksum, K. (1988). Estimation and testing in the two-sample generalized odds-rate model. *Journal of the American Statistical Association*, 83:744–749.

Dai, H. and Bao, Y. (2009). An inverse probability weighted estimator for the bivariate distribution function under right censoring. *Statistics and Probability Letters*, 79:1789–1797.

Analysis for Time-to-Event Data under Censoring and Truncation.
http://dx.doi.org/10.1016/B978-0-12-805480-2.50008-3,

Dai, H. and Fu, B. (2012). A polar coordinate transformation for estimating bivariate survival functions with randomly censored and truncated data. *Journal of Statistical Planning and Inference*, 142:248–262.

Diggle, P. J., Heagerty, P., Liang, K.-Y., and Zeger, S. L. (2002). *Analysis of Longitudinal Data*. Oxford University Press, 2nd edition.

Ding, J. and Wang, J.-L. (2008). Modeling longitudinal data with nonparametric multiplicative random effects jointly with survival data. *Biometics*, 64:546–556.

Dore, G. J., Freeman, A. J., Law, M., and Kaldor, J. M. (2002). Is severe liver disease a common outcome for people with chronic hepatitis c. *Journal of Gastroenterology and Hepatology*, 17:423–430.

Feinleib, M. (1960). A method of analyzing log-normally distributed survival data with incomplete follow-up. *Journal of the American Statistical Association*, 55(291):534–545.

Fine, J., Ying, Z., and Wei, L. (1998). On the linear transformation model with censored data. *Biometrika*, 85:980–986.

Fleming, T. R. and Harrington, D. P. (1991). *Counting processes and survival analysis*. John Wiley & Sons.

Fleming, T. R., Harrington, D. P., and O'Sullivon, M. (1987). Supremum versions of the logrank and generalized wilcoxon statistics. *JASA*, 82:312–320.

Freeman, A., Dore, G., Law, M., Thorpe, M., Overbeck, J., Lloyd, A., Marinos, G., and Kaldor, J. M. (2001). Estimating progression to cirrhosis in chronic hepatitis c virus infection. *Hepatology*, 34:809–816.

Fu, B., Tom, B. D., Delahooke, T., Alexander, G. J., and Bird, S. M. (2007). Event-biased referral can distort estimation of hepatitis c virus progression rate to cirrhosis and of prognostic influences. *Journal of Clinical Epidemiology*, 60:1140–1148.

Gijbels, I. and Gürler, U. (1998). Covariance function of a bivariate distribution function estimator for left truncated and right censored data. *Statistica Sinica*, 1998:1219–1232.

Gilbels, I. and Wang, J. (1993). Strong representations of the survival function estimator for truncated and censored data with applications. *Journal of Multivariate Analysis*, 47:210–229.

Gross, S. T. and Lai, T. L. (1996). Nonparametric estimation and regression analysis with left-truncated and right-censored data. *Journal of American Statistical Association*, 91:1166–1180.

Gürler, U. (1996). Bivariate estimation with right truncated data. *Journal of American Statistical Association*, 91:1152–1165.

Gürler, U. (1997). Bivariate distribution and hazard functions when a component is randomly truncated. *Journal of Multivariate Analysis*, 60:20–47.

Haller, B., Schmidt, G., and Ulm, K. (2013). Applying competing risks regression models: an overview. *Lifetime Data Analysis*, 19:33–58.

He, S. and Wong, X. (2003). The central limit theorem of linear regression model under right censorship. *Science in China Series A*, 46:600–610.

He, S. and Yang, G. (1998). Estimation of the truncation probability in the random truncation model. *The Annals of Statistics*, 26(3):1011–1027.

He, S. and Yang, G. (2000). The strong uniform consistency of the p-l estimator under left truncation and right censoring. *Statistics and Probability Letters*, 49:235–244.

He, S. and Yang, G. (2003). Estimation of regression parameters with left truncated data. *Journal of Statistical Planning and Inference*, 117:99–122.

Henderson, R., Diggle, P., and Dobson, A. (2000). Joint modelling of longitudinal measurements and event time data. *Biostatistics*, 1:465–480.

Horner, R. D. (1987). Age at onset of Alzheimer's disease: clue to the relative importance of etiologic factors? *American Journal of Epidemiology*, pages 409 – 414.

Hougaard, P. (1986). A class of multivariate failure time distributions. *Biometika*, 73:671–678.

Hougaard, P. (2000). *Analysis of Multivariate Survival Data*. Springer-Verlag, New York, first edition.

Huang, J., Vieland, V., and Wang, K. (2001). Nonparametric estimation of marginal distributions under bivariate truncation with application to testing for age-of-onset anticipation. *Statistica Sinica*, 11:1047–1068.

Huber, P. J. (1973). Robust regression: asymptotics, conjectures and Monte Carlo. *The Annals of Statistics*, 1(5):799–821.

Hyde, J. (1980). Survival analysis with incomplete observations. *Biostatistics Casebook*, pages 31–46.

Kalbfleisch, J. and Prentice, R. (2002). *The Statistical Analysis of Failure Time Data*. John Wiley & Sons, New York, second edition.

Kaplan, E. and Meier, P. (1958). Nonparametric estimation from incomplete observations. *Journal of the American Statistical Association*, 53:457–481.

Keiding, N. and Gill, R. (1990). Random truncation models and markov processes. *The Annals of Statistics*, 18(2):582–602.

Kim, J. H. (1993). Chi-square goodness-of-fit tests for randomly censored data. *The Annals of Statistics*, 21.

Klein, J. and Moeschberger, M. (2003). *Survival Analysis Techniques for Censored and Truncated Data*. Springer.

Koul, H., Susarla, V., and Van, R. (1981). Regression analysis with randomly right censored data. *The Annals of Statistics*, 9:1276–1288.

Lagakos, S. W., Barraj, L. M., and Degruttola, V. (1988). Nonparametric analysis of truncated survival data, with application to aids. *jBiometrika*, 75:515–523.

Lai, T. L. and Ying, Z. (1991a). Estimating a distribution function with truncated and censored data. *The Annals of Statistics*, 19(1):417–442.

Lai, T. L. and Ying, Z. (1991b). Large sample theory of a modified Buckley-James estimator for regression analysis with censored data. *The Annals of Statistics*, 19(3):1370–1402.

Lai, T. L. and Ying, Z. (1991c). Rank regression methods for left-truncated and right-censored data. *The Annals of Statistics*, 19(2):531–556.

Lai, T. L. and Ying, Z. (1994). A missing information principle and m-estimators in regression analysis with censored and truncated data. *The Annals of Statistics*, 22(3):1222–1255.

Lin, D. and Ying, Z. (1993). A simple nonparametric estimator of the bivariate survival function under univariate censoring. *Biometrika*, 80:572–581.

McCullagh, P. and Nelder, J. A. (1989). *Generalized linear models (Second edition)*. London: Chapman & Hall.

Mei-Cheng Wang, Nicholas P. Jewell, W.-Y. T. (1986). Asymptotic properties of the product limit estimate under random truncation. *The Annals of Statistics*, 14(4):1597–1605.

Merzbach, E. and Nualart, D. (1988). A martingale approach to point processes in the plane. *The Annals of Probability*, 16:265–274.

Miller, R. G. (1976). Least square regression with censored data. *Biometrika*, 63:449–464.

Miller, R. G. and Harplen, J. (1982). Regression with censored data. *Biometrika*, 69:521–531.

Murphy, S., Rossini, A., and van der Vaart, A. (1997). Maximum likelihood estimation in the proportional odds model. *Journal of the American Statistical Association*, 92:968–976.

Murtaugh, P. A., Dickson, E. R., Dam, G. M. V., Malinchoc, M., Grambsch, P. M., Langworthy, A. L., and Gips, C. H. (1994). Primary biliary cirrhosis: prediction of short-term survival based on repeated patient visits. *Hepatology*, 20:126–134.

Oakes, D. (1989). Bivariate survival models induced by frailties. *JASA*, 84(406):487–493.

P. K. Bhattacharya, H. C. and Yang, S. S. (1983). Nonparametric estimation of the slope of a truncated regression. *The Annals of Statistics*, 11(2):505–514.

Pettitt, A. (1984). Proportional odds model for survival data and estimates using ranks. *Applied Statistics*, 33:169–175.

Prentice, R., Moodie, F., and Wu, J. (2004). Hazard-based nonparametric survivor function estimation. *Journal of Royal Statistical Society*, B(66):305–319.

Rezza, G., Lazzarin, A., Angarano, G., Sinicco, A., Pristera, R., Tirelli, U., Salassa, B., Ricchi, E., Aiuti, F., and Menniti-lppolito, F. (1989). Tje natural history of hiv infection in intravenous drug users: Risk of disease progression in a cohort of serconverters. *AIDS*, 3:8790.

Ritov, Y. (1990). Estimation in a linear regression model with censored data. *The Annals of Statistics*, 18:303–328.

Serfling, R. J. (1980). *Approximation Theorems of Mathematical Statistics*. John Wiley & Sons.

Sharma, N. K. and Sherker, A. H. (2009). Epidemiology, risk factors, and natural history of chronic hepatitis C. *Clinical Gastroenterology*, pages 33–70. edited by K. Shetty and G.Y. Wu.

Shen, P. (2006). An inverse-probability-weighted approach to estimation of the bivariate survival function under left-truncation and right-censoring. *Journal of Statistical Planning and Inference*, 136:4365–4384.

Shen, P. (2015). Parameter estimation in regression for long-term survival rate from left-truncated and right-censored data. *Communications in Statistics - Simulation and Computation*, 44(4):958–978.

Stute, W. (1993). Consistent estimation under random censorship when covariables are present. *Journal of Multivariate Analysis*, 45:89–103.

Stute, W. (1996). Distributional convergence under random censorship when covariables are present. *Scandinavian Journal of Statistics*, 23(4):461–471.

Su, Y.-R. and Wang, J.-L. (2012). Modeling left-truncated and right-censored survival data with longitudinal covariates. *The Annals of Statistics*, 40(3):1465–1488.

Sweeting, M., Angelis, D. D., Neal, K., Ramsay, M., Irving, W., and et al., M. W. (2006). Estimated progression rates in three United Kingdom hepatitis C cohorts differed according to method of recruitment. *Journal of Clinical Epidemiology*, 59:144–152.

Therneau, T. M., Grambsch, P. M., and Fleming, T. R. (1990). Martingale-based residuals for survival models. *Biometrika*, (73):147–160.

Tisiatis, A. A. and Davidian, M. (2001). A semiparametric estimator for the proportional hazards model with longitudinal covariates measured with error. *Biometrika*, 88:447–458.

Tisiatis, A. A. and Davidian, M. (2004). Joint modeling of longitudinal and time-to-event data: an overview. *Statistica Sinica*, 14:809–834.

Tsai, W., Leurgan, S., and Crowley, J. (1990). Nonparametric estimation of a bivariate survival function in the presence of censoring. *The Annals of Statistics*, 14:1351–1365.

Tsai, W.-Y., Jewell, N. P., and Wang, M.-C. (1987). A note on the product-limit estimator under right censoring and left truncation. *Biometrika*, 74(4):883–886.

Turnbull, B. W. (1976). The empirical distribution function with arbitrarily grouped, censored and truncated data. *Journal of the Royal Statistical Society. Series B (Methodological)*, 38(3):290–295.

van den Berg, G. J. and Drepper, B. (2016). Inference for shared-frailty survival models with left-truncated data. *Econometric Reviews*, 35(6):1075–1098.

van der Laan, M. (1996a). Efficient estimation in the bivariate censoring model and repairing NPMLE. *The Annals of Statistics*, 24:596–627.

van der Laan, M. (1996b). Nonparametric estimation of the bivariate survival function with truncated data. *Journal of Multivariate Analysis*, 58:107–131.

Wang, H., Dai, H., and Fu, B. (2013). Accelerated failure time models for censored survival data under referral bias. *Biostatistics*, 14:313–326.

Wang, M.-C. (1991). Nonparametric estimation from cross-sectional survival data. *Journal of the American Statistical Association*, 86(413):130–143.

Woodroofe, M. (1985). Estimating a distribution function with truncated data. *The Annals of Statistics*, 13:163–177.

Woolson, R. F. (1981). Rank tests and a one-sample log rank test for comparing observed survival data to a standard population. *Biometrics*, 37:687–696.

Yu, B. and Peng, Y. (2008). Mixture cure models for multivariate survival data. *Computational Statistics and Data Analysis*, 52:1524–1532.

Index

Printed in the United States
By Bookmasters